Country on the Move:
Migration to and within Israel, 1948–1995

The GeoJournal Library

Volume 42

The titles published in this series are listed at the end of this volume.

Country on the Move: Migration to and within Israel, 1948–1995

by

GABRIEL LIPSHITZ
Department of Geography,
Bar-Ilan University,
Ramat-Gan, Israel

KLUWER ACADEMIC PUBLISHERS
DORDRECHT / BOSTON / LONDON

In cooperation with the Jerusalem Institute for Israel Studies

A C.I.P. Catalogue record for this book is available from the Library of Congress.

ISBN 0-7923-4850-8

Published by Kluwer Academic Publishers,
P.O. Box 17, 3300 AA Dordrecht, The Netherlands.

Sold and distributed in the U.S.A. and Canada
by Kluwer Academic Publishers,
101 Philip Drive, Norwell, MA 02061, U.S.A.

In all other countries, sold and distributed
by Kluwer Academic Publishers Group,
P.O. Box 322, 3300 AH Dordrecht, The Netherlands.

Printed on acid-free paper

Printed in the Netherlands

To my parents, who immigrated in 1935 and, under difficult conditions, put their energy and resources into their children's education in Israel

CONTENTS

TABLES

FIGURES

HISTORICAL OVERVIEW

Between the establishment of the State of Israel in 1948 and the end of 1995, roughly 2.4 million people immigrated to Israel. They include approximately one million immigrants who arrived in the 1950s, mostly from North Africa and Asia, and around 700,000 who arrived in 1990–1995, most of them from the former Soviet Union (FSU).

Quantitatively, immigration to Israel has not been gradual; instead, it has been marked by very large numbers of immigrants, known as "waves of immigration," in short periods of time (see fig. 1). At the time of independence in 1948, Israel had a Jewish population of 717,000; the wave of immigration in the 1950s more than doubled the figure. Immigrants who arrived in the 1990s currently account for approximately 15 percent of the total Israeli Jewish population. These two values are exceptional in global immigration patterns. In this sense, immigration to Israel is a "shock of change" (a term coined by the renowned economist Gunnar Myrdal in 1957), with an impact on the economic, social, settlement, and cultural structure of the country, not only on the national level, but notably on the regional—geographical—level, too.

In the 1960s and 1970s, and even more so in the 1950s, the Israeli government intervened directly and vigorously in all aspects of immigrant absorption. New towns

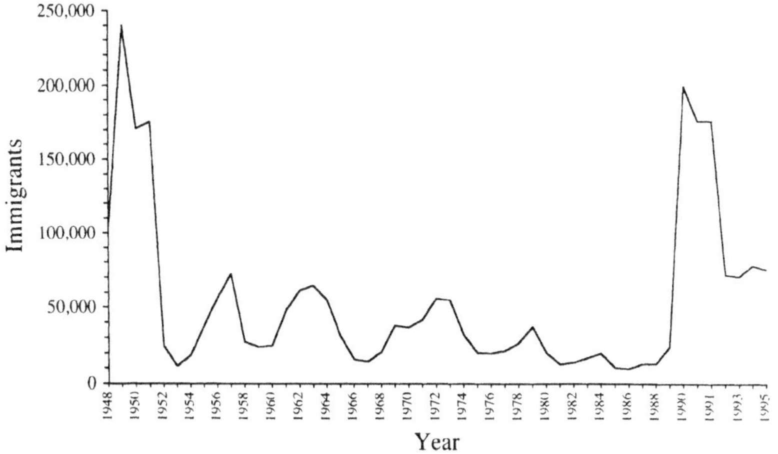

Fig. 1: Immigration to Israel, 1948–1995

and rural settlements were built and new factories were established, mainly in the national peripheral regions—the Galilee and the Negev—to give the immigrants homes and jobs. In other words, in the 1950s the Israeli government did not let *market forces* absorb the immigrants, disperse them spatially, and provide them with jobs. "*Planning from above*" dominated immigrant absorption as well as other aspects of Israeli economic life.

In the early 1960s immigration to Israel declined dramatically. The "trickle" that arrived in the 1960s, 1970s, and 1980s had very little effect on the population distribution and the distribution of economic activity in Israel. *During these decades, a different type of population movement shaped the country's spatial pattern: internal migration between regions within Israel by immigrants and nonimmigrants alike.* Globally, interregional migration is a clear spatial result of socioeconomic gaps between regions. As a rule, people move from regions with a low "level of development"—the periphery—to prosperous regions—the core. In most countries, including Israel, interregional migration is a definite manifestation of the spatial flow of "market forces."

In late 1989 another substantial wave of immigrants began to arrive in Israel. The wave intensified in the early 1990s, and by 1995 it had brought 700,000 newcomers to Israel, the vast majority of them from the former Soviet Union. During this period, approximately 30,000 Ethiopian immigrants also arrived, including 14,000 in an impressive one-day operation ("Operation Solomon") initiated by the Israeli government with assistance from other countries, most importantly the United States. This book does not address the experience of the Ethiopian immigrants, which requires an anthropological research method more than a spatial-economic one.

The absorption of immigrants from the FSU in the 1990s has been governed by a completely new policy conception, known in Israel as *direct absorption.* This policy is an about-face in thinking regarding immigrant absorption. Under the new policy, the government gives every immigrant, upon arrival in Israel, an "absorption package" that includes a sum of money meant to cover expenses for a limited period; the immigrants may do with this sum as they see fit in all areas of life: housing, employment, education, consumption, and so on. The direct-absorption track thus lets the immigrants' own wishes and personal considerations determine the course of their absorption. The declared intent of the government policy was to reduce the immigrants' dependence on national and municipal government; to make the absorption process more efficient and flexible; and to speed up the immigrants' integration in housing, employment, and social and cultural life. In other words, the Israeli government decided to switch *from a strategy of intervention from above to a strategy of absorption by market forces.* It should be noted that this change is not limited to Israeli immigrant-absorption policy; rather, it reflects a change that has been taking place in the economic policy of the Israeli government for almost 20 years, namely, a gradual transition from a (relatively) centralized economic regime to a free-market regime. When the volume of immigration from the FSU to Israel increased, however—far beyond the forecasts—the government partially abandoned planning of absorption by market forces and began to intervene to a certain degree. This intervention chiefly affected housing, as the government built thousands of dwelling units, most of them in the national periphery, where there were large reserves of publicly-owned land.

The goal of the present book is to describe and analyze the spatial-economic-demographic results of immigration and internal migration in Israel between 1948 and 1995. As stated above, there were two peaks of immigration during this period: the wave of the 1950s, during which the Israeli government intervened vigorously ("planning from above") to disperse the immigrants geographically; and the wave of the 1990s, when the government adopted a policy of absorption by market forces ("planning from below") to disperse the immigrants. Between these two peaks, the main population flow determining the spatial makeup was internal migration.

The waves of immigration to Israel are just a few of many *flows of international migration. Internal migration*, too, is an extremely salient geographical phenomenon in the spatial makeup of every country. The spatial patterns of these flows, the demographic and economic traits of the immigrants, government policy in the host countries, the economic impact of the immigrants on the countries of origin and destination and on specific regions in these countries, and the immigrants' impact on the development of the settlement network vary with time. The geographical directions of internal migration also change over time, thereby contributing not only to a change in the physical spatial pattern, but also to changes in the socioeconomic spatial pattern. In what ways does immigration to Israel resemble other flows of international migration and in what ways does it differ? For example, can we learn anything about the policy adopted in Israel from studying the economic or social immigrant-absorption policies of other countries? Where do the immigrants tend to settle initially, and what is the spatial pattern of their redistribution in Israel and other countries? Does the economic impact of immigrants on the national and regional levels in Israel resemble that in other countries? To answer such questions, a survey was done of the professional literature, taking a look at flows of international migration from several standpoints. This survey makes up the bulk of *chapter 1* and provides the global and theoretical context of the study on the following subjects: the geographical distribution of immigrants on the global level; the geographical distribution of immigrants within a country; internal migration by immigrants and the consequent internal migration by natives (though this is also affected by other factors); the economic integration of immigrants in the destination countries and their impact on the economy of the countries of origin and destination; and immigration policy. The literature survey will focus on internal migration, attempting to understand the role it plays in reducing demographic and economic disparities between the core and peripheral regions and the impact of these disparities on patterns of internal migration. These issues form the background for various topics that are examined in the book.

One immigration wave that had a marked effect on the geographical distribution of the Israeli population and of economic activity was that of the 1950s. The main considerations in adopting policy measures for the geographic dispersion of the immigrants of the 1950s were security-based, geopolitical, and economic. Application of this policy was made possible by the arrival of one million destitute immigrants, who were at the mercy of the government with respect to housing and jobs. What policy measures were adopted to disperse the immigrants geographically in the 1950s? What were the spatial results of this policy? In other words, to what extent did the population-dispersion policy reduce the polarity of the population distribution between core and peripheral regions? What were the spatial-economic results of this policy? Have the economic and social disparities between the core and peripheral regions increased or decreased as a result of the population-dispersion policy? These questions are the subject of *chapter 2.*

As we shall see, the "spatial encounter" between the pre-State (pre-1948) immigrants, who were predominantly from Europe, and the immigrants of the 1950s, who came mostly from North Africa and Asia, was responsible for creating socioeconomic disparities in Israel. These disparities were manifested in the formation of a spatial-socioeconomic structure of core and periphery. Internal migration since the early 1960s was one of the spatial results of these disparities, and was the main demographic factor in shaping the spatial physical and social patterns in Israel. *Chapter 3* describes the internal-migration trends as regards the spatial core-periphery structure in the 1960s, 1970s, and 1980s. Because this migration was strongly affected by socioeconomic disparities between Israeli regions, and because these disparities were a spatial result of the waves of immigration, a special section of chapter 3 is devoted to describing regional disparities in Israel.

Chapter 4 examines the *geographical distribution of the immigrants of the 1990s on the macro level: periphery versus core.* It begins with a spatial-policy dilemma concerning the housing market: should the government have encouraged the immigrants to settle in central Israel, where job prospects are good, while offering incentives to promote long-range suburbanization toward the national periphery by affluent nonimmigrants (who can commute to employment centers in central Israel)?[1] Or should it have acted as it did in the 1950s and used the immigrants to populate the periphery, in an effort to kick-start the periphery out of its demographic and economic stagnation? The chapter then proceeds from the dilemmas to application, looking at *actual housing-market policy*, including the geographical distribution of public residential building versus private residential building (the geographic expression of supply). The last part of the chapter explores the spatial results of government policy and of market forces with respect to the geographical distribution of the immigrants. In other words, it looks at the geographic housing preferences of the immigrants, including their internal migration from their initial place of settlement (the geographic expression of demand).

Chapter 5 looks at spatial-economic aspects of the absorption of immigrants from the FSU, and especially at employment. It begins with a description of the macro-level (national-level) employment dilemma: the gap between the supply of educated *human capital* with scientific and academic professions on the one hand and the demand for and utilization of this human capital on the other. It also considers the changes over time and space in the employment of immigrants the longer they have been in Israel, and the changes over time in the ratio of immigrant unemployment rates to the rates among nonimmigrants, by gender and level of education. The chapter then moves from the national level to the geographical/regional level, describing and analyzing the high quality of human capital of the immigrants who settled in towns in the peripheral regions of Israel, their integration in the local job market, and the utilization of their human capital for the economic growth of the localities. The chapter concludes by looking at small businesses and factories as sources of jobs for immigrants in peripheral development towns.

1. Suburbanization among affluent nonimmigrant Israelis is a spatial process that began in the late 1970.

The concluding section of the book is divided into two parts. The first part—which forms the bulk of the conclusion—is devoted to a comprehensive discussion of the findings of the study and their significance for spatial immigrant-absorption policy. The second part analyzes the findings in the theoretical context and compares the trends in Israel with international-migration trends elsewhere in the world, which are described in the chapter 1.

The book is based on numerous, diverse sources. The main ones are as follows: studies and papers from the professional literature; publications by the Central Bureau of Statistics, the Ministry of Immigrant Absorption, the Ministry of Construction and Housing, and Local Government Economic Services; publications and internal reports by local authorities; surveys in five development towns of nonimmigrants and immigrants, factory owners, owners of workshops, and owners of small businesses; and interviews with high-ranking officials, including mayors and local-council heads, town engineers, directors of absorption departments, directors of Employment Service offices, and directors of strategic-planning and economic-planning units.

This book is, in part, the integrated product of several studies that I carried out in the past few years, but also presents, for the first time, the results of a comprehensive study on the absorption of immigrants from the former Soviet Union.

I would like to express my profound thanks to the following people who contributed directly and indirectly to the book: Prof. Arie Shachar, director of the Institute for Urban and Regional Studies at the Hebrew University of Jerusalem; Prof. Yehuda Gradus, director of the Negev Center for Regional Development at Ben-Gurion University of the Negev; Prof. Aharon Kellerman, vice-president of the University of Haifa; Prof. Michael Sonis of Bar-Ilan University; Prof. Rami Friedman, head of the Jerusalem Institute for Israel Studies; and Dr. Daniel Felsenstein of the Hebrew University. I would also like to thank the staff of the Jerusalem Institute for Israel Studies for its administrative and technical assistance, and especially Ora Ahimeir, the Institute director; my research assistants, Tali Gvirtzman, Noa Feller, Noa Lipshitz, and Nati Tabursky; the people who conducted the polls in the development towns; the immigrants who answered the survey questions; Orly Haimi of the Bar-Ilan University Cartography Laboratory, who prepared the illustrations and maps; Tammy Sofer and Michal Kidron of the Hebrew University, who prepared a few of the figures; Deborah Limmer, who translated the book from Hebrew; and Lenn Schramm, who did the typesetting.

While writing this book, I was awarded two research grants related to its preparation: from the GIF (the German-Israel Foundation), for research on the core-periphery aspects of immigration from the former Soviet Union; and from the Jerusalem Institute for Israel Studies, for research on the distribution of immigrants from the former Soviet Union in Israeli development towns.

Last but not least, I would like to thank my wife, Devora, and my children, Noa, Yair, Netanel, and Tamar, for creating the atmosphere that makes my academic work a pleasure.

To all, I extend my sincerest thanks.

Gabriel Lipshitz
Jerusalem, August 1997

CHAPTER 1

IMMIGRATION: AN INTEGRATIVE PERSPECTIVE

1.1 Spatial Patterns of Immigration: The International Level

Worldwide political and economic changes have altered the spatial patterns of international migration (Salt 1989; Castels and Miller 1993; King 1993b; Nijkamp and Spiess 1993, 1994a, 1994b; Gould and Findlay 1994). The *political world order* during most of the second half of this century revolved around the Cold War between the United States and the USSR and these two superpowers' control of various parts of the world. When the Communist regime collapsed, bringing down the Iron Curtain between Eastern and Western Europe, Western European and other Western countries experienced new influxes of immigration, requiring special measures by the destination countries. *Immigration from the former Soviet Union to Israel is one outcome of the recent worldwide political changes.*

The *economic world order* in the second half of this century is linked to the growth of multinational corporations and the new geographical division of the worldwide labor force. The transfer of the production lines of multinational corporations to developing countries should lessen emigration of cheap labor from these countries to the mass-production lines of the industrialized countries. However, while production lines were moving to developing countries, a "brain drain" began, as highly skilled workers from the Third World moved to developed countries. The emigration of young, educated, skilled workers from the developing countries to the developed countries has been one of the most prominent characteristics of international migration in the last few decades. Such migration increases the economic disparity between core and peripheral countries, which in turn induces even more migration (Gould and Findlay 1994). Emigration of a skilled and educated work force from the developing countries detracts from these countries' potential to develop modern industries that pay relatively high salaries. On the other hand, these migrants have the cumulative effect of enhancing human capital in the developed countries and speeding up the development of high-technology industries. It should be noted that despite the relatively large proportion of young, educated migrants from developing to developed countries, many members of the cheap labor force also migrate each year, in the hope of bettering their financial situation in the developed countries.

The worldwide spatial pattern of international migration in the 1980s and 1990s consists of movement of migrants from the developing countries, where rates of natural increase are high, to the developed, industrialized countries, where rates of natural increase have dropped markedly. For example, natural increase in the countries of the Common Market/European Community declined from 8.0 per 1,000 population in 1960 to 1.8 per 1,000 population in 1989; in the same period net immigration rose from 0.2 to 3.3 per 1,000 population (Hall 1993). Differences in rates of natural increase produce a labor surplus and crowded conditions in the villages and cities of the developing countries and a labor shortage in the developed countries. Migration between them is a demographic-economic manifestation of these disparities.

The Third World countries are growing at a rate of 3 to 4 percent a year, and their big cities are growing at an annual rate of 5 percent. Natural increase is the main cause of growth in developing countries, although their big cities are also growing due to internal migration from the villages and small towns. The developing countries are exporters of migrants, and immigration is not a major factor in their growth. In contrast, the populations of developed countries such as Germany, the United States, and Canada would hardly increase at all without immigration (Öberg and Boubnova 1993). Until the 1950s Western Europe was a significant source of emigration, but since then—for the past four decades—these countries have had positive immigration balances, with immigration exceeding emigration (Öberg 1993).

Within this worldwide spatial pattern, the *regional dimension* is of particular importance. Hall argues that much of international migration is "regional" and geographically short-range. For example, most immigrants to Germany are from southern and eastern Europe; most immigrants to France are from North Africa, especially Algeria. The geographic pattern of immigration in Europe is chiefly from the peripheral countries in the south to the core countries in the west and, within the core countries, to the big cities (King 1993a). The United States attracts a large number of immigrants from nearby Mexico, though also from the more distant Philippines and South Korea (Hall 1993). The rate of immigration to Australia from Southeast Asia has increased considerably over time. These examples illustrate Hall's argument regarding the regional nature of immigration patterns. Immigration from Asia to Western Europe and North America, as we shall see below, confuses the regional dimension of international migration by expanding its range. In contrast to Hall, Portes and Rumbaut argue that most immigration today is long-range, and that distance is becoming less of a consideration (Portes and Rumbaut 1990).

Immigration to Israel from the former Soviet Union is an example of long-range, rather than regional, immigration. It is motivated by political, national, and religious factors. Under the Law of Return, enacted in 1950, shortly after Israel gained its independence (1948), all Jews in the world are entitled to move to Israel, irrespective of their financial status, health, and age.[1] Moreover, since 1948 the Israeli government has adopted various policy means with regard to housing, employment, education, and consumption to support immigrants to Israel.

Significant changes have taken place over time in spatial patterns (countries of origin and destination) and in the profile of the immigrants (demographic and economic traits). Western European destination countries were fed first by Italy, then by Spain, Greece, and Portugal, later by Yugoslavia and Turkey, and eventually by North Africa and the Third World (King 1993c). Salt (1992a, 1992b), in his paper on types of international migration, discusses the spatial changes that have occurred since the middle of this century: In the 1940s migration was mainly related to World War II and the events

1. Although the Law of Return allows for exceptions in cases of applicants likely to endanger the public health, this provision is customarily ignored.

that followed. The 1950s saw the resumption of permanent international migration, particularly the movement of Europeans to North America, Australia, and New Zealand, and the first trickle of workers migrating to Western Europe. Migration between Eastern and Western Europe was blocked by the Iron Curtain, which separated the countries in the Soviet sphere of influence from the West. The 1960s were marked by migration of workers due to the efforts of Western European countries to develop and rehabilitate their economies and the consequent labor shortage. Migration in the 1970s was characterized chiefly by reunification of workers' families, primarily in Europe. During this period other Western countries began to tighten up their immigration policies, including their policies regarding the admission of temporary immigrants and foreign workers, for fear of family reunification and permanent settlement, as was happening in Europe. Because of the immigration of guest workers to Europe (intended to be temporary) and the ensuing family reunification, Europe now has roughly five million guest workers. In the 1980s illegal immigrants, political-asylum seekers, and educated, skilled workers joined the flow of international migration. In his analysis of changes in the types of immigration to Europe over time, White states that Western Europe experienced three major waves of immigration (shown in fig. 2): (1) a wave of labor immigration; (2) a wave of family reunification; and (3) the post-industrial wave (White 1993).

The wave of labor migration to Europe began in the 1950s and continued until the 1970s. It was a reaction to the economic requirements of the destination countries, which needed workers, especially for unskilled labor, to fuel its economic growth in the 1950s and 1960s. The countries of origin of the immigrants to Europe have varied over time: first the colonies and former colonies, then southern Europe, and then North Africa and the Far East. The second wave, according to White, is the family-reunification wave, which began in the mid-1970s and lasted about ten years (White 1993).

This second wave shows how reality overrides planning. The original plan was

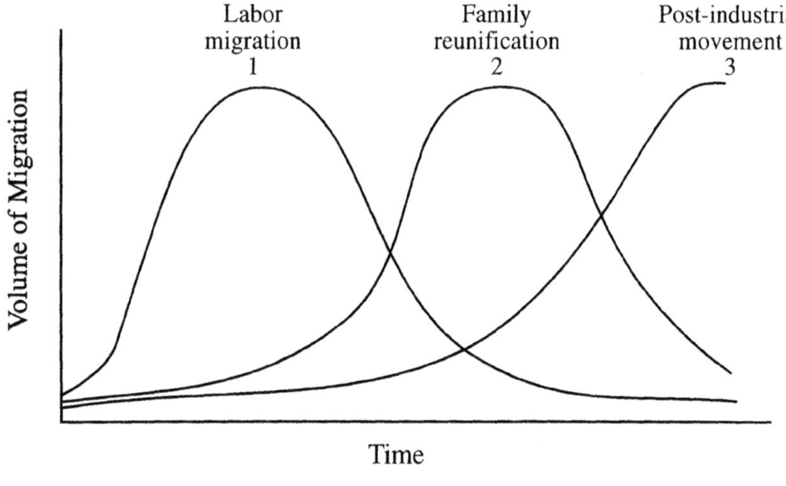

Fig. 2: Three waves of international migration (based on White 1993)

that the immigrants (predominantly men) to Western Europe would stay temporarily and would help the economies of the host countries recover from World War II. The temporary immigrants became permanent. Pressure from families who had remained in the countries of origin to reunite with the spouse (generally the husband) who had migrated to Western Europe was heavy. The process of family reunification, which began in the 1970s, added hundreds of thousands of immigrants to Western European countries.

The third wave is connected to a structural change in the European economy: from traditional industry to sophisticated, high-tech industry, business services, and computer services. This wave has three distinct components, as shown in figure 1: high-caliber professional workers, whose migration is a manifestation of the change in the economic structure of Western Europe; illegal immigrants; and political-asylum seekers. All these groups existed before, but their numbers increased in the 1980s.

The changes in the spatial patterns of immigration include not only changes in the countries of origin and in the profile of the immigrants, but also changes in the destination countries. Whereas in the 1960s and 1970s the migrants' main destinations were Western Europe, the United States, and Canada, in the 1980s the oil-rich countries of the Middle East took the place of Europe.

International migration in the 1980s was essentially made up of a group of *spatial networks* with similarities and differences: for example, migration from North Africa to Europe; from Eastern Europe to Western Europe; from the Caribbean islands and Mexico to the United States; from various African countries to South Africa and to some countries on the African coast; and from Southeast Asia to Australia; as well as migration of workers from Southeast Asia to the oil-rich Middle Eastern countries.

Salt's analysis of *international migration networks* demonstrates that there are differences between and within the networks, and that each network changes over time (Salt 1992b). For example, Germany and France have fairly similar numbers of immigrants and rates of immigration, but from different places. Immigrants to England come from still other countries. In the 1960s Italians stopped moving to France and turned to West Germany instead due to wage differentials and the return from Algeria to France of about one million French people.

The Middle Eastern network (migration to the oil-rich countries) differs from the European one not only in terms of the countries of origin but also in immigration policy; most of the migrants are from Arab countries, and family reunification is prevented. Unlike in Western Europe and North America, most immigrants are channeled into certain areas of work, chiefly simple manufacturing and services (Birks, Seccombe, and Sinclair 1988). When oil prices fell in the mid-1980s, the rulers of the countries that employed foreign workers replaced workers from Arab countries with Asians, who are more efficient and cheaper, and thus the composition of migration in the Middle Eastern network changed (Addleton 1991). In 1975, approximately 65 percent of foreign workers in the oil countries were from Arab countries; by 1985 only 30 percent were. In the latter year approximately 63 percent were from southern Asia (Pakistan, Bangladesh, India) and eastern Asia (the Philippines, Indonesia, Thailand). The Gulf War caused the departure of many Arab workers, as Palestinians were expelled from Kuwait and Egyptians were expelled from Iraq; their place was taken by workers from southern and eastern Asia.

Until the 1970s the immigration network in Australia was characterized by immigration from Europe; this is long-range immigration, and in this respect it is not part of a spatial network. In the 1980s the percentage of immigrants from Southeast Asia increased. Australia's strong economic links with Great Britain, Europe, and the United States have grown weaker, whereas its ties with Asia—especially Southeast Asia—are becoming stronger (Chalkey and Winchester 1991). Between 1970 and 1991 the share of Australian exports to Asia rose from 49 percent to 67 percent, while the share of exports to Western Europe fell from 23 to 15 percent. The Australian establishment stopped treating Asia like a military and demographic threat and began to regard it as an economic opportunity. The changes in regional economic ties are also manifested in immigration to Australia—51 percent of immigrants to Australia in 1991–1992 were Asian—and therefore immigration from Asia to Australia can be expected to increase over the years.

Immigration to Canada and the United States has undergone a trend similar to that seen in Australia. In the 1940s and 1950s, 70 percent of immigrants to Canada were from Europe; by the 1980s the figure was down to about 20 percent. Asians, who constituted 10 percent of the immigrants to Canada in the 1940s and 1950s, accounted for 60 percent in the 1980s. In 1960 about 40 percent of all immigrants to the United States were from Europe; in the 1980s only 10 percent were. In the United States, too, the Europeans were replaced by southern and eastern Asians, who accounted for about 14 percent of all immigrants to the United States in the 1960s, as against 40 percent in the 1980s. Although the trends in the US, Canada, and Australia are similar in the sense that the proportion of Asian immigrants has increased, there is a difference. Whereas immigration by Asians (specifically, from Southeast Asia) to Australia is part of the emerging spatial-economic network of Southeast Asia, which includes Australia, migration by Asians to the US and Canada is an aspect of long-range international migration processes.

In the 1990s a new region of attraction began to emerge: eastern Asia. Over the years the Far Eastern countries have altered their economic structure, and in the 1990s they have some of the highest growth rates in the world. A study by the World Bank shows that in the 1980s the economic growth rates of 23 eastern Asian countries exceeded those of all other countries. Particularly strong growth was seen in the "Four Tigers" (Hong Kong, South Korea, Singapore, and Taiwan), in the three countries known as the Newly Industrializing Economies, or NIE (Indonesia, Malaysia, and Thailand),[2] and in China. These eight countries are also known as High Performing Asian Economies (HPAE). One interesting finding of the World Bank study is that the economic growth was not accompanied by an increase in inequality; in several countries income disparities (according to the GINI and other indices) actually narrowed (World Bank 1993). The rapid economic growth caused a major improvement in human welfare. Life expectancy rose significantly in these countries, from 56 in 1960 to 71 in 1990. The percentage of people living in poverty (without sufficient basic living conditions such as clean water, food, and shelter) fell dramatically: from 58 percent in 1960 to 17 percent in 1990 in Indonesia, and from 37 to 5 percent in Malaysia.

2. These countries are also known as the NIC (Newly Industrializing Countries).

Two main factors are behind this impressive economic growth: investments by private entrepreneurs and investments in human capital. As a result of the rapid economic growth and the improvement in human welfare, these countries ceased to be only "exporters" of migrants and became "importers." They now attract people not only from southern and eastern Asia, but from Western Europe and North America as well.

As described above, the *distribution of immigration* in the 1980s and 1990s increasingly reflects the *globalization of the economy*—structural changes in the global economy (transition from traditional industry to high-tech industry and business services) and geographical changes in unemployment rates. The main reasons for this are the growth of high-tech industry, the expansion of global commerce, and the expansion of multinational companies that require educated, professional employees (Findlay and Garrick 1990; Kanjanapan 1995). Because such workers are not always available in the local market, the United States, Canada, and Australia have placed great emphasis since the 1980s on admitting "high-caliber" immigrants (Salt and Ford 1993). The professional job market is currently the motivating force behind the global economy, in which the importance of sophisticated industry is increasing (Findlay 1990; Kanjanapan 1995; Findlay et al. 1996). When there is a disparity between supply and demand of such workers in the local market, companies make sure to narrow the gap by importing workers in the international labor market.

Demand for workers also depends on unemployment levels: immigration decreases as the unemployment rate rises (King 1993c). Because unemployment in Europe is fairly high, we might expect to see a decrease in the import of workers (and a drop in immigration). However, a shortage of professional workers in Europe works against this trend. This incongruity is connected to the structural change in the Western European economy, manifested in a rise in demand for professional workers at the expense of non-professionals. In Britain, for example, experts predict a 27 percent rise in demand for scientists and engineers and a 19 percent drop in demand for industrial and agricultural workers (Salt and Ford 1993). Consequently, destination countries seek professional, managerial, and technical (PMT) workers suited to the changes in their economic structure and their comparative advantage in the world market, and they have reduced their demand for unskilled workers.

In light of the structural changes in the Western economy, Salt foresees that the developed countries will continue to attract high-caliber professional workers, whereas the need for non-professional workers will lessen over time. The developed countries will encourage emigration by non-professionals and will strive to keep their professional workers. The structural changes in the global economy, reflected in the decline of industry and the rise of business services and computer services, have impaired the job prospects of non-professional immigrants. Immigration by professionals depends on the individual field and on compatible economic development in the developed countries. Despite these changes, there is still some demand for non-professional workers, but it is declining (Salt 1992a).

So far we have described global economic changes that affect the worldwide spatial pattern of migration. In this context we should note two east Asian countries: Japan and China. The two countries have vastly different economic structures. In the past few years China has undergone profound changes in its economy and in its economic relations with the global market. However, the worldwide patterns of supply and demand will be affected considerably by the policies of China and Japan. China may have the greatest potential of all to supply (i.e., to "export") emigrants. It is estimated that, as of 1992, 130 million people in the rural sector of China were unemployed, as were 30 million in the urban sector. If the Chinese government were to liberalize its emigration policy, the world could be immediately inundated with Chinese workers. It should be noted that, in the destination countries, the Chinese are considered excellent entrepreneurs and efficient workers. Thus, China's economic and emigration policies may have a dramatic quantitative impact on the supply of migrants in the world market.

Japan is an example of a country that blocks demand for foreign labor, even though the high standard of living and economic development in Japan have created a labor shortage. In the early 1990s there were 35 percent more jobs available than job-seekers. In particular, there is a shortage of skilled workers. Japan continues to administer a stringent policy regarding the admission of foreign workers, and a change in this policy would unquestionably have an impact on the worldwide pattern of migration (Stahl and Appleyard 1992). Thus, a change in the Japanese immigration policy could have a substantial impact on the worldwide spatial pattern of demand for migrants.

As described above, the worldwide structural changes in the economy have altered the pattern of demand for labor. The volume and spatial pattern of international migration, however, will be determined by the extent to which the local market of each country can produce workers on a high enough technological and scientific level to meet the demand. The wider the gap is between supply and demand in the high-tech market, the more international migration of educated and skilled workers there will be. In Europe, for example, the prevailing view is that skilled workers will have to be imported because of an extremely low rate of natural increase (Coleman 1992). The low rates of natural increase in Western Europe are causing a cumulative shortage of labor, including a shortage of educated and skilled workers. Nevertheless, the increase in the percentage of women in the labor force may well reduce the demand for foreign workers, including skilled ones. Coleman estimates the potential of the female labor force in Europe at 33 million workers, and he calls this the "hidden supply." Another source of labor within Europe is Eastern Europe, which has a supply of skilled and professional workers. The last source of educated and skilled workers in Europe consists of the Western European countries themselves, where the unemployment rate among the university-educated is high. There are currently about 15 million unemployed persons in Western Europe, including scientists and skilled workers; as a result, international migration in Europe may become predominantly intra-European rather than global (Coleman 1992). Indeed, the new restrictions on immigration, the disintegration of the Soviet regime, the collapse of the Iron Curtain, the reunification of Germany, and the substantial increase in free movement among EU countries support a new pattern of international migration in Europe—more European and less international.

Despite these trends, it can be expected that in certain segments of the economy Europe will experience immigration of expert workers that the European countries are unable to obtain locally or from elsewhere in Europe. The increase in modern industry and business services requires a high degree of skill that at times can be achieved only by importing workers (Findlay and Garrick 1990).

Due to changes in spatial patterns and composition, international migration in the 1990s can be expected to widen worldwide interregional disparities. Most destination countries are becoming more selective; they are most interested in educated people with scientific and technological occupations. The era of mass migration of unskilled workers has ended. As Salt writes, "Their need for mass immigration is gone and will not return" (Salt 1992b, 1081).

The 1990s also represent the end of a period of fairly easy immigration and the beginning of an era of *strict immigration policies* (Salt 1992a). The popular destination countries are becoming more stringent and selective. This change affects not only the spatial patterns of international migration, but also its composition. People with prestigious occupations in high technology and other fields are permitted to immigrate to attractive destinations. Thus we are likely to see two types of immigration patterns: legal immigration of young people with higher education and academic and technological occupations and illegal immigration of unskilled laborers.

To conclude the discussion of global spatial patterns, it is worth presenting Sassen's research findings. Sassen, who stresses the impact of global economic processes, argues that in addition to the internationalization of trade, which was a significant factor in 19th-century immigration, the internationalization of investment and production has become a major factor in the late twentieth century. The worldwide spatial distribution of international investment and production has a major impact on the spatial pattern of international migration (Sassen 1988, 1993). These global processes, Sassen believes, develop over time. In the mid-1960s investment in the developing countries by North American and Western European countries and Japan expanded, and as a result manufacturing for export in the developing countries picked up. In the 1970s, and even more so in the 1980s, the major cities in the West became important economic nodes in the management and control of economic systems. The main change in the 1990s has been that the United States has become a main center for investment, after having been a major exporter of investments in the previous 30 years. These economic changes are having decisive influences on international migration, as is described below.

Sassen does not argue that the traditional factors—poverty, population density, high birth rates, and economic decline—have ceased to stimulate international migration (especially from developing to developed countries). She does believe, however, that they are becoming proportionately less important in the explanation, as the global division of the organization of production and economic activity becomes much more substantial.

The structural changes in the North American, Western European, and Japanese economies caused a shortage of unskilled labor, which was then "imported" from the developing countries (less so in Japan). Most immigrants to these countries start out as industrial production workers and as simple laborers in the services, although in the past few years immigration by academics, scientists, and skilled workers has increased substantially.

One of the effects of investment in developing countries—some of which focuses on the manufacture of high-technology products—is enhancement of their human capital (for a comprehensive discussion relating to Hong Kong, see Findlay 1990; Kanjanapan 1995; Findlay et al. 1996). This enhancement is due to changes in the factories themselves, generally induced by the government of the country in question. Thus, over time the quality of the work force in developing countries improves. Some of these workers do not find jobs to suit their skills, and they migrate to Western countries. Consequently, Sassen notes, additional countries start to export migrants (those with professional expertise and higher education); some examples are Singapore, Malaysia, and Indonesia (Sassen 1993).

The increase in the GNP of the immigrants' countries of origin (the developing countries) is chiefly due to greater external investment by most Western countries. A large portion of these investments is directed toward export industries. The main reason why Western countries have chosen to invest in developing countries is that the large, inexpensive work force in these countries enables multinational corporations to lower the prices of their products (because they pay workers low wages) and to compete effectively in the global market. Sassen argues that the increase in employment in several developing countries as a result of investment by the developed countries is accompanied by a simultaneous increase in export of migrants. Thus, there appears to be a two-way link between the rise in the growth rates in the economies of certain developing countries and the rise in their rates of emigration. In effect, this is a chain reaction that begins with investment and culminates in migration. Investment in the developing countries by Western entrepreneurs has substantially boosted production in these countries, including the manufacture of high-tech products. These changes have altered the demand in the local work force, which now needs more engineers, more scientists, and more people in academic and technical fields. The local market (aided by government policy) has adapted itself to the new demand and has begun to "produce" an educated, skilled work force, some of which is a potential reserve for migration to the West.

In agriculture, there is a logical explanation for this relationship. The increase in agricultural output has to do with adoption of modern production methods and intensified use of agricultural machinery. As a result, many workers lost their jobs and migrated to the major cities in the developing countries and other countries. However, the growth in industrial output is a result of substantial investment in labor-intensive factories (some of them high-technology factories). Why, then, does emigration increase?

Sassen does not provide a clear answer; instead she calls for more research on the subject. However, her book suggests that the explanation has to do with the cumulative benefit to the caliber of the work force in a small number of developing countries. Sassen notes that the United States, Canada, Japan, West Germany, Britain, France, and the Netherlands are responsible for 90 percent of investments in developing countries, and that over half of these investments are in five or six countries. This work force, whether because of economic and political ties with Western businesses or because of its job skills, finds its way to Western countries. The changes in the caliber of the work force in the developing countries is caused, inter alia, by the expansion of investments in their *export industries*. Export industries naturally boost the economic ties between the populations of the investing countries and the developing countries. The latter become familiar with the global market and learn to suit their occupational skills to the changes in the global market. Consequently, more and more workers from developing countries are "absorbable," occupationally speaking, in Western countries.

The structural changes in the global economy and the economic ties that develop between Western countries (those that invest and import immigrants) and a small number of developing countries (the recipients of the investments and exporters of emigrants) are the chief factors behind international migration at the end of this century. Many poor developing countries with high unemployment and a traditional economic structure are not represented among the main providers of international migrants.

It appears that immigration to Israel from the former Soviet Union is neither part of an international migration network nor caused by the globalization of the economy. It is primarily the result of a political and economic push and a national-religious pull (the Law of Return).

1.2 The Geographic Distribution of Immigrants in the Destination Countries

Above we described changes in the spatial patterns of international migration as a result of geopolitical, geo-economic, and geodemographic changes on the global level. What are the spatial patterns of immigrants *within* the destination countries? The present section consider the answer to this question, based on examples from the distribution of immigrants within the United States.

The most significant factor explaining the geographical distribution of immigrants in the destination countries at the turn of the century was geographical proximity to the country of origin (Portes and Rumbaut 1990). It is no coincidence that, at the start of the twentieth century, European immigrants settled on the east coast of the United States, whereas the Asians settled in California and along the rest of the west coast. The same factors can explain the concentration of Mexicans in the southwestern United States. Portes and Rumbaut attribute the importance of geographical distance to economic and geographic considerations: the lower cost of migration and return migration; and familiarity with the land and with climatic conditions (in the Mexican case).

Geography—proximity to the country of origin and/or to the seacoast where the immigrants' ships landed in the destination country—was the dominant factor behind the distribution of immigrants in the destination country in the 19th and early 20th centuries. This factor is more relevant to the initial immigrants, whereas subsequent waves preferred to live near the earlier arrivals so as to be close to members of their own ethnic group. This geographical proximity reflects the immigrants' need for cultural, social, psychological, and even financial support from their families, friends, and members of their ethnic group.

Later, economics became a factor in the geographical distribution of immigrants in the United States and in most other destination countries. Communities of immigrants moved farther inland due to the growth of the metal and automobile industries in the Midwest, as well as the development of railroads. Throughout the periods of immigration to the United States, cultural-ethnic familiarity was an important consideration in the immigrants' choice of destination. In other words, the first immigrants from a particular country, settled for various reasons in a particular place, and their choice of location influenced that of later immigrants from the same country. The first immigrants provide economic and sociocultural support for the subsequent ones, and they form the basis of the information network that immigrants need even before arriving in the destination country.

Over the years the geographical importance of proximity to the country of origin has decreased, and economic factors have become more important. For example, there are two different geographical patterns of immigration in the United States in the late twentieth century: concentration and dispersion (Portes and Rumbaut 1990). In other words, although immigrants live in every state and every big city, only a few states and a few big cities have major concentrations. For example, Portes and Rumbaut note that in 1987, 71 percent of legal immigrants to the United States settled in six states and about 50 percent in just three states: California, New York, and Florida. Sassen notes that in the 1990s, due to massive migration by Mexicans, California is the leading state in terms of immigrant absorption. New York and California take in about half the immigrants. New Jersey, Illinois, Florida, and Texas receive about 25 percent. This concentrated pattern has an immediate impact on the housing and employment markets, as well as on the demographic-cultural structure of these states.

The patterns of concentration and dispersion also depend on the immigrants' countries of origin: most Cuban immigrants in the United States live in Miami, those from the Dominican Republic are concentrated in New York, and the Koreans are concentrated in Los Angeles; the Indians, on the other hand, are more widely dispersed. Some of the concentration of immigrants (for instance, the Cubans and the Vietnamese) is due to US government policy. In general, Portes and Rumbaut (1990) argue that when the US government has intervened in the settlement of immigrants, the result has been a highly concentrated geographical pattern, whereas when (as in the majority of cases) the government has not intervened the pattern has been more dispersed.

Immigrants to the United States are geographically concentrated on the level of cities as well as states. In general, immigration to the United States in the past two decades can be regarded increasingly as immigration to the major cities. In 1987 only 7 percent of legal immigrants settled in non-urban areas, while over 50 percent settled in ten cities. Illegal immigrants may have a greater tendency to move to rural areas, and from there—after improving their economic situation—to the cities (Portes and Rumbaut 1990). The main factor that attracts immigrants to the big cities is economic and employment-related. Even if many of them—including professionals—find work at the bottom of the job ladder, they expect to improve their job status and subsequently their income, and therefore they favor the big cities as their destination.

This geographical concentration of immigrants applies mainly to the first and second generations. Presumably, their children will migrate within the country, creating a pattern of dispersion. The disappearance of the concentrated pattern among groups of immigrants thus reflects their cultural assimilation within the host society. The more rapid this process is, the more quickly they will disperse geographically. Some term this process the "succession theory": the first arrivals (or their children) assimilate and leave the urban centers to newer arrivals. In contrast to this theory of geographical dispersion of immigrants as a function of assimilation, one can propose a theory that supports the continued existence of ethnic centers—even if they move to other parts of the big cities (Portes and Rumbaut 1990). This concentration is essential to the individual—even if he or she has climbed the socioeconomic ladder—for social, cultural, and political reasons.

As mentioned above, immigrants tend to settle initially in the major cities, and within the cities in specific neighborhoods, usually poor inner-city neighborhoods (Simon 1992). This tendency was particularly prominent in White's first and second waves of immigration to Europe (the labor-immigration and family-reunification waves), as described in the beginning of this chapter. In these neighborhoods the new immigrants find people who arrived in previous waves of immigration from the same country of origin. The reason for the concentration of immigrants in the big cities has to do with the good job prospects there; sociocultural and economic considerations are responsible for their concentration in specific neighborhoods.

Relatives and friends give the newcomers psychological, social, and even financial support and guide them through the halls of bureaucracy (Pekin 1989; Lipshitz 1992). How long immigrants remain dependent on their relatives is a function of their advancement on the socioeconomic ladder of the host society. The more rapidly they advance, the shorter the period of dependence on relatives is. In most cases the desire to live in geographical proximity also fades (Pekin 1989), and a certain period of time after their initial settlement those who are doing well financially begin to move to other neighborhoods. Their choice of place of residence is based no longer on ethnic affiliation, but on socioeconomic status.

The spatial patterns set by the first immigrants from a particular ethnic group therefore have a major impact on later immigrants from the same ethnic group. There are even cases in which residents of a town in a certain country of origin settled in a particular town in the destination country and were followed by further migration from the same town of origin to the same destination town (Lipshitz 1992). This has been termed "chain migration" (King 1993c).

Immigrant entrepreneurs who deal in products and services consumed largely by members of their own ethnic group also tend to settle in or near the large ethnic concentration. For them this concentration is an important source of business and a basis for the development of economic ties (Portes and Rumbaut 1990). These spatial patterns hold true more for blue-collar workers and traditional entrepreneurs than for white-collar workers, who tend to spread out even at first.

White shows that in the big European cities foreigners have increased their share of the population over time, while hardly spreading at all into the rural areas and the small and medium-sized cities. This process is caused by the arrival of immigrants and the departure of natives. For example, foreigners accounted for approximately 21 percent of the population of Amsterdam in 1990, about 17 percent of the population of London in 1991, and about 24 percent of the population of Frankfurt in 1991 (White 1993).

After analyzing several studies, White reaches the general conclusion that immigrants do not disperse at random within cities. Their geographical location within a city has a lot to do with their inferior economic position and their consequent inability to pay high rents. Affluent immigrants, who generally account for a relatively small proportion of the newcomers, choose high-class neighborhoods; some examples are the Japanese community in Dusseldorf and the employees of international organizations in Brussels (Glebe 1986). The maps in White's paper show clearly that the foreign population in London is concentrated in something resembling an arc around the inner-city neighborhoods. The picture in Paris is similar, except that in Paris minorities also live along a line stretching from the St.-Denis quarter to the city center. This axis coincides with the traditional industrial zone of Paris (White 1993).

The spatial distribution of the immigrants' initial place of residence within the cities determines the later spatial distribution of immigrants, too. In London, for example, immigration created geographic concentrations in the heart of the city. About 25 percent of the residents of the inner city of London are minorities (Fielding 1993b). Nevertheless, as immigrants and their children assimilate in the host society, they begin to disperse. Along with the geographic dispersal and social assimilation, a "pushing" process is at work: the newcomers are pushed into segregated ethnic neighborhoods and they push out the previous round of immigrants or their children (Portes and Rumbaut 1990). In Chicago, for example, this process has led, in less than a century, to a dramatic change in the geographical distribution of minorities in the city.

On the national level, too, the spatial distribution of the earliest immigrants determines the spatial distribution of subsequent immigrants. A good example of this is the spatial distribution of Hispanic immigrants in the United States (Davis, Haub, and Willette 1983). Although Hispanics live in almost every US state today, most still live in a few states and regions. California, with a population of 6.5 million Hispanics, is the initial place of settlement for one out of every three Hispanic immigrants. Other states with large Hispanic populations are Texas, with 4.1 million Hispanics; New York, with 2.1 million; and Florida, with 1.5 million. Hispanics are also geographically segregated by ethnic group: 73 percent of Mexicans in the United States live in California, Florida, and Texas; 61 percent of Puerto Ricans in the United States live in New York and New Jersey; and 59 percent of Cubans in the United States live in Florida. This spatial distribution seems to be determined chiefly by the geographical proximity of the countries of origin to US states. Moreover, 88 percent of the Hispanics in the United States live in metropolitan areas, and more than half of these live in the central cities of those metropolitan areas.

This geographical proximity is the initial cause of the spatial segregation of immigrants along ethnic lines, and it applies chiefly to the early waves of immigration. However, the ethnic-spatial segregation continues and even intensifies, because immigrants wish to live near others from their country of origin, with whom they feel more secure. This sociocultural environment facilitates integration and alleviates the psychological tensions that affect many immigrants in their first few years in the country.

The immigrants, Hall says, move to the big cities for the same reasons their relatives and others moved there one hundred years earlier (Hall 1993). The reasons for immigration to London, New York, Berlin, and other cities, as in the past, are economic, employment-related (high probability of finding a job of some sort), and sociocultural (proximity to relatives and community services). In other words, Hall sees the geographical concentration of the immigrants in the destination countries as the outcome of a combination of economic and sociocultural factors. The big cities in the West are apparently destined to be "host cities" for the people of the developing world. Immigration is turning the big cities in the West into an economic bridge to the Third World. Vancouver is an interesting example. For many years Vancouver has been a regional economic center for western Canada, especially for the province of British

Columbia. In the past decade it has undergone a significant spatial-economic transformation from a regional business center to an international business center linking Canada with eastern and southern Asia, including Japan. One of the main factors in this change is immigration to Vancouver, which accelerated greatly in the past decade. The increasingly affluent immigrants in Vancouver are an important human basis for the creation and expansion of economic ties with business centers in their countries of origin.

In some countries the destination regions of immigrants vary over time. Canada is a good example. In the 1940s and 1950s, and even in the 1960s, the main destination regions were the peripheral provinces, especially Manitoba and Saskatchewan. In the 1980s, in contrast, the main destinations were the core provinces: British Columbia and Ontario. In 1911 and 1921, about 40 percent of immigrants to Canada moved to the prairie provinces and 30 percent to Ontario. In 1986 the figures were 15 and 55 percent, respectively. This change reflects the structural changes in the Canadian regional economy: a rapid transition from agriculture to industry and from industry to business services in Ontario and the western provinces (Moore, Ray, and Rosenberg 1990).

Changes over time in immigrants' choice of area of residence occur on the *intra-urban level*, too. In the first wave of migration to Europe (labor immigration; see White 1993), which took place in the 1950s and 1960s, the immigrants—who came mainly for work—settled in the big industrial cities, particularly in the poor neighborhoods. Some of these neighborhoods are located in the inner city and others in suburbs on a low socioeconomic level. Suburban residential neighborhoods for immigrants are found most often in countries in which ethnic minorities are eligible for public housing. For example, in Amsterdam and Paris in the 1970s a rapid process of deconcentration of immigrants from the inner city to the suburbs began after public housing was built for them (Sporton and White 1989; van Amersfoort and de Klerk 1987). In Britain, the deconcentration of immigrants was not as conspicuous as in the Netherlands and France, because in some British cities public housing was built for immigrants in the inner city. The second wave of the 1970s (family reunification) reinforced the spatial pattern of the 1950s and 1960s. The third (post-industrial) wave in the 1990s, which comprises mainly high-caliber professionals, is concentrated on the national level in cities with high-technology industry and business services and in the relatively affluent neighborhoods of these cities; whereas refugees and illegal aliens continue to settle in the poor neighborhoods of the big industrial cities (White 1993).

1.3 Internal Migration (Redistribution of Natives and Immigrants)

Immigration to specific destinations affects *internal migration* to and from these destinations. In Champion's opinion, this is becoming one of the most important issues in the study of international migration, although relatively speaking—particularly in Europe—it has not been given the attention it deserves (Champion 1996). The geographical distribution of whites and minorities in the United States in the 1980s was strongly affected by the interplay between immigration and internal migration (Frey 1995a, 1995b). In US states and metropolitan areas that attract immigrants—especially Hispanic and Asian minorities—net internal migration by whites is negative and net internal migration by minorities is positive. These spatial processes function simultaneously in a "scissor action": large numbers of immigrants move in and large numbers of whites move out. Together they constitute the main factor shaping the spatial distribution of the US population on the national and metropolitan levels. California, New York, and Texas are examples of states whose demographic-spatial makeup is determined by international migration (the entry of minorities and the resultant departure of whites), whereas the demography of Arizona, Virginia, and North Carolina is determined mainly by internal migration of whites (positive net migration by the white population).

This dynamic process—which picked up speed in the 1980s and operates in just a few US states—leads to what Frey (1995b, 1996) has called a "racial and ethnic balkanization pattern." Interestingly, according to Frey the departing whites are those with low levels of income and education. These natives are displaced in the job market by immigrants. This process occurs with great intensity in US metropolitan areas. The metropolitan areas in which immigrants play an important positive role in population growth and natives play a negative role through internal migration (negative net migration) are the old and long-established ones, such as Los Angeles, Chicago, and New York. In relatively new metropolitan areas, such as Atlanta, Seattle, and Phoenix, the main factor affecting population growth is internal migration.

While the immigrants tend to concentrate in the big cities—and especially in the inner cities—the native-born begin to migrate from the inner cities to the outskirts and the suburbs. In Hall's opinion, one effect of immigration has been a long-term decline of the large Western cities (Hall 1993). Hall believes that the stage identified by some researchers as reconcentration of population in the big cities is merely a statistical error or a misinterpretation of spatial processes. The big Western cities are in a stage of super-dispersal or super-deconcentration, accompanied by reconcentration of population in the medium-sized cities located in the distant periphery of the big cities and not in the big cities themselves. The bulk of this process is carried out through internal migration by the native-born, whereas immigrants are responsible for the growth of the big cities (the central cities of the metropolitan areas) in the West (Hall 1993).

What factors do immigrants take into consideration when deciding to leave their initial place of settlement and migrate within the destination country? An examination of internal migration by immigrants within Canada (Nogle 1994) indicates that higher education and proficiency in the language—which is important for gathering information about housing and employment—increase their tendency to migrate internally. Furthermore, the longer an immigrant remains in the destination country, the more dominant economic considerations become compared to linguistic considerations in decisions regarding internal migration. The desire for access to family and friends remains an important consideration in the decision to migrate, especially for people whose economic situation is poor. Another finding is that the immigrants' initial place of settlement has an impact on whether they choose to migrate within the destination country. Those who initially settled in the major cities—Toronto, Montreal, and Vancouver—are less likely to move to other cities.

This brief survey of the spatial redistribution and socioeconomic traits of the immigrants shows that immigrants are increasingly moving to the big cities while natives leave these cities for the metropolitan periphery and medium-sized cities. In addition, migration between Western cities has been dropping, and immigration from developing countries to Western countries has increased markedly. Stringent immigration policies enable Western countries to be selective as to the quality of the immigrants. As a result, in some destination countries the immigrants' socioeconomic and occupational levels are higher than the average in that country. The immigrants tend to move from their initial place of settlement, and a certain percentage of them move to regions based more on their socioeconomic status than on their ethnocultural affiliation. These processes alter the internal sociodemographic and cultural structure of the big cities and reshape the spatial distribution of the population on the national and metropolitan levels.

Immigration is only one of the factors that affect internal migration by natives within a country. In fact, internal migration between regions in a country operates even during periods when the volume of immigration is relatively small. *Internal migration is partly—and perhaps primarily—a spatial result of socioeconomic disparities between regions.* In some countries, including Israel, the socioeconomic disparities between regions are themselves a spatial result of the geographical distribution of immigrants and their socioeconomic attributes. Internal migration includes natives as well as immigrants who first settled in a particular region and moved elsewhere, most often for economic reasons. Both quantitatively and qualitatively (in terms of the demographic, social, and economic traits of the migrants), the movement of population between regions is a very significant factor in shaping the social, economic, and physical spatial patterns of any country.

Over time, does internal migration strengthen core regions demographically and economically and weaken peripheral regions, or does it work in the opposite direction? Do changes take place in the geographical direction of migration, from the national (and metropolitan) periphery to the core, from the core to the periphery, and again from the periphery to the core? A theoretical debate began on this question in the 1970s, and became even more prominent in the early 1980s. The focal question was whether the changing pattern of net migration in core and peripheral regions of developed and developing countries is moving in the direction of convergence or divergence.[3] In other words, has the positive migration balance in the core areas increased in positivity or decreased, even to the point of turning negative, over time? Conversely, has the negative migration balance in the national peripheries become more or less negative over the years, or has it turned positive?

Discussions of this topic started to appear in various studies in the 1970s (Morrison and Wheeler 1976; Vining and Strauss 1977; Klassen and Paelinck 1979; Vining and Kontuly 1978). In the 1980s the discussion took the form of a debate among

3. This book focuses on "large" geographical units: core and periphery. Although the constancy of boundaries between units and changes in the boundaries over time are extremely relevant in discussions of the core-periphery structure of internal migration, they will not be discussed here. On this subject, see Fielding 1982, Champion 1988, and Lipshitz 1996b. All of these papers argue that in research on the core-periphery structure of internal migration one should at least distinguish between migration from the core to the periphery of the core (suburbanization) and migration from the core to the national periphery (internal migration, usually accompanied by a change in place of employment).

the various schools of thought. Research conducted in the 1980s was to provide a basis for the cohesion of various approaches and schools of thought within the context of population distribution. The study that first brought up the argument about the change in net migration patterns in core and peripheral areas was Vining and Pallone 1982.

Vining and Pallone (1982) show how the positive migration balance in the core regions of Western countries has declined, while the negative migration balance in the peripheries of the same countries has become less negative. Their overall conclusion is that "in the developed world at least, the century-long migration toward the high-density core regions is over" (p. 339). In net-migration terms, this implies a converging spatial process in developed countries. Vining and Pallone show a completely different spatial pattern in developing countries, where the positive migration balance in the core has remained stable, and in some countries has become increasingly positive. The opposite is true of the peripheries of these countries, where the negative migration balance has either remained stable or become more negative. In net-migration terms, this implies a diverging spatial pattern.

According to Cochrane and Vining (1988) the 1980s were a turning point in the spatial migration patterns of developed countries. These patterns are different from those described in the Vining and Pallone studies. Cochrane and Vining found that in most Western countries in the 1980s, net migration in the core regions became more positive, while that in the periphery became more negative. The change is the result of increased migration from the periphery to the core. Signs of a strengthening core and spatial concentration can also be found in such non-Western countries as Korea and Taiwan. These findings brought Cochrane and Vining to the conclusion that the 1980s produced a decline in "counter-urbanization," or "rural renaissance" (Gordon 1979), and a renewal of the urbanization process, as defined by Fielding (1982): a direct link between net migration and settlement size. According to Cochrane and Vining the spatial pattern of net migration in the 1980s changed direction from convergence (from core to periphery) to divergence (from periphery to core).

Mera (1988) analyzes spatial processes in Japan of the 1980s and identifies a process of renewed concentration among populations in the central metropolitan areas. He interprets the trends identified by Cochrane and Vining as "cyclical movements in migration," but he does not agree with this definition. According to Mera, "Metropolitanization is not a unidirectional growth path as it used to be thought, nor is it a life cycle of mere growth and decline. It is more likely to occur of growth and decline" (p. 270). Thus, according to Mera the spatial trend of population dispersion is from convergence to divergence. Mera notes that this conclusion is more suited to developed countries. He gives three reasons for the changes in the Japanese migration patterns: (a) spatial distribution of public investment; (b) the concentration of information-processing activities; (c) liberalization of the financial sector. The last change brought about the rapid growth of banks and private commercial companies in metropolitan areas.

These studies discuss the quantitative, rather than the qualitative, aspects of spatial migration. A series of studies published by Brown and his colleagues, which analyzes the social and economic characteristics of the migrants, are an important contribution to this field (Brown and Goetz 1987; Brown, England, and Goetz 1989; Brown and Lawson 1989). The study on Venezuela (Brown and Lawson 1989) reveals signs of population dispersion to the periphery in qualitative as well as quantitative terms since the beginning of the 1970s. This process alters the quality of human capital in the periphery. When migrants to the peripheral regions are highly educated and have academic and scientific professions, they enhance the quality of human capital in these regions, thereby creating a potential for the development of high-technology industries; for such companies, human capital is the most important locational factor.

It should be noted that internal migration is not only a factor in reducing or increasing the demographic disparities between core and periphery, as we saw above. It is also a major factor in reducing or increasing the economic disparities between core and periphery (for a comparative analysis, see Lipshitz 1992b). One of the main points in the theoretical dispute between the spatial-equilibrium and spatial-disequilibrium schools has to do with the effect of internal migration on economic disparities between regions. The spatial-equilibrium school argues that migration from the periphery to the core—due to income differentials—reduces marginal output in the core and increases it in the periphery, thereby reducing the disparities between them in marginal output and per-capita income. In other words, internal migration acts to reduce economic disparities between regions—and government intervention is unnecessary.

On the other hand, the spatial-disequilibrium school maintains that internal migration is a major factor in increasing regional disparities, and only government intervention can reduce them. As they see it, because the migrants from poorer areas (periphery) to flourishing areas (core) include a high percentage of educated and skilled persons, such migration detracts from the economic growth of the periphery and cumulatively contributes to the economic growth of the core.

International and internal migration are thus two significant population flows that affect physical and economic spatial patterns in many countries. In some countries they work in the same geographical direction, strengthening the core regions; in other countries they weaken the core regions and strengthen the periphery of the core, or even the national periphery.

1.4 The Economic Factor in Immigrant Absorption: The National and Regional Levels

The chief factor underlying international migration is still the economic situation in the countries of origin and destination. People migrate in the expectation of finding work (suiting their job skills) and in the hope that their income level will improve substantially. International migration affects the economic development of both the country of origin and the destination country. Fielding (1993a) divides the economic factor in the destination countries into three levels:

LEVEL 1: THE ROLE OF THE BUSINESS CYCLE

Economic production will increase in a particular country at a particular time due to demand for a product or products manufactured in that country. The increase in sales boosts profits, thereby enhancing the confidence of businesspeople. Business owners then invest more, and among the effects of the greater investment are additional job openings that boost the demand for labor. Some of this demand for labor is filled by immigrants. Economic history shows that the business cycle is made up of alternating slumps and periods of prosperity. The slumps begin with a cumulative process of economic crisis and drop in sales and investments, followed by a decline in the number of jobs available and, consequently, a drop in the volume of immigration to that country; even the volume of return migration to the country declines.

The business cycle thus affects the volume of immigration and produces cycles in the housing market. There are three intertwined circles: migration, business, and housing. The waves of immigration to Germany in good years and emigration during the crises of 1967 and 1974/75 are a good example of this (Böhning 1979). Immigration to Germany while the country was experiencing economic growth boosted the housing market, and these two factors combined caused a further demand for workers, which was met, inter alia, by immigration.

The business cycle and its connection to international migration and domestic economic processes can be divided into four stages (Fielding 1993a). First is the period of economic growth: The immigration rate increases and net immigration is positive. Housing prices rise and unemployment drops. The second stage is the peak of the cycle: Immigration is at its highest point, and emigration is at its lowest. Housing prices are at their peak, and unemployment is down. Net immigration is extremely positive. In the third stage—the period of economic decline—immigration begins at the peak and falls rapidly, while emigration increases rapidly. Net immigration is negative, housing prices drop, and unemployment rises. In the fourth stage, the trough, immigration is at its lowest point and emigration is at its peak. Housing prices are low, unemployment is high, and net immigration is extremely negative.

Level 2: The role of transformations in the organization of production in mass migration

Most studies on the link between immigration and economic development have focused on the business cycle, which acts on the economy in the short term. Much of the research has ignored longer-term economic changes, which are gradual and cumulative. The longer-term changes are transformations in the organization of production, known today as "restructuring." The concept of restructuring says that there are cumulative changes in the worldwide spatial division of production, and consequently in the spatial division of labor (Massey 1984). The 1960s, 1970s, and 1980s were a time of cumulative spatial structural changes in the global economy: corporate managements, especially those of multinational firms, were attracted to the major cities of Europe (London, Paris, etc.) and North America (New York, Chicago, Los Angeles, Toronto, etc.); research and development units were drawn to areas with comfortable weather and environmental conditions and a supportive academic environment; and factories moved to places where labor was cheap, especially the peripheral European countries and the Third World.

Fielding (1993b) and Salt (1989) argue that this process can be expected to reduce international migration by unskilled workers (in whose countries of origin manufacturing is located) and increase migration of educated and skilled workers. Indeed, the major cities of Western Europe have lost a large share of their manufacturing jobs as a result of global deindustrialization, which has reduced demand for the unskilled labor generally supplied by immigrants. The transformation of these cities, some of which are known as "global cities," into management and R&D centers has attracted a new breed of immigrant: scientists, managers, and businesspeople (see Shachar and Öberg 1990).

These gradual and cumulative global changes were disrupted in 1989–1992 by a great shock: the collapse of the Communist bloc. This fall simultaneously exposed the economic disparities between Western and Eastern Europe and made emigration from east to west possible. Consequently, while global economic processes might be expected to reduce the volume of migration, and especially mass migration, to Western Europe, the fall of the Iron Curtain opened up a tremendous potential for migration (King 1993a).

LEVEL 3: THE ROLE OF "DEEP STRUCTURAL" PROCESSES IN MIGRATION

Economic developments over centuries created two major blocs: developed countries and developing countries. These economic disparities cannot be eliminated from one decade to the next. They cause the population of the Third World to aspire to improve its standard of living, and one way of doing so is to migrate to Western countries. Fielding shows that these profound gaps are an economic-historical cause of continuing migration to Western Europe (Fielding 1993a).

Thus, a variety of economic processes influence migration trends. However, international migration affects economic processes, too, via the influx of new labor, new entrepreneurs, and new investments, and even by means of societal and cultural changes that take place in the destination countries and cities due to migration. The minority populations of the big cities sometimes serve as an "economic bridge" for owners of businesses in their countries of origin, and in this way they affect the economy of the destination region.

1.5 The Economic Integration of Immigrants and Their Impact on the Economy of the Destination Country

Two main questions are of interest to researchers who study the economic integration of immigrants: First, how do the immigrants adjust to the destination country and assimilate in its economy? Second, what is the effect of international migration on the economies of the countries of origin and destination? Regarding the first question, Borjas argues that no single theory can provide a complete answer, although the human-capital theory is one of the most important in the field and therefore many scholars use elements of it to examine the integration of immigrants in the economy of the destination country (Borjas 1989). Chiswick, too, stresses in several studies that the immigrants' adjustment to the economy and societal structure of the destination country depends mainly on the quality of their human capital (Chiswick 1978). For the most part, immigrants are deficient in education, language, geographic proximity to jobs, and other qualities that are valuable in the destination country. Over time, they learn the language, invest in education and a trade or profession, and migrate internally within the country, thereby adjusting to the geo-economic structure of the country. The enhancement of their human capital is manifested in a reduction of the earnings gap between immigrants and natives. According to Chiswick, the immigrants' investment in human capital depends on the environment from which they come (attitudes toward higher education, etc., in their country of origin), their age, and the length of time they expect to remain in the destination country.

Another factor that affects immigrants' economic integration in the destination country is the length of time they have been there. By means of an economic analysis and graphic illustration Borjas shows that, in the same immigration cohort, people in their forties earn higher salaries than do people in their thirties, and people in their thirties earn more than those in their twenties. Furthermore, people in their forties (and in their thirties and twenties) who immigrated in the 1960s (the 1960s cohort) earn more than those who immigrated in the 1970s, who in turn earn more than people who immigrated in the 1980s (Borjas 1985).

Immigration can affect several aspects of a destination country's economy: local demand for goods and services, demand for production factors, demand for physical capital (e.g., housing), economies of scale and agglomeration economies, wages, unemployment, inflation, balance of payments, national exports, and internal migration of the work force (Greenwood 1994). Below we survey some of these areas of influence:

Immigration of workers can affect production and wages in the destination regions for better or for worse. This effect is discussed in the professional literature from two standpoints: that of *production theory* and that of *public-sector expenditure*. In the former, scholars examine whether immigrants and natives complement or replace each other as production factors. In the latter, the researchers are interested in the ratio of the immigrants' contribution to state or local government revenues through the taxes they pay to the cost of their consumption of welfare and other public services.

Greenwood examines the effect of immigration on production and wages in the destination country (Greenwood 1994). By analyzing a series of graphs pertaining to the United States and the rest of the world (see fig. 3), Greenwood shows that when immigration to the United States does not affect the labor-demand curve (the demand curve D_{US} remains as it was prior to immigration, rather than shifting outward), *wages fall* from W_{US} to W'_{US} due to the downward shift in the supply of labor from S_{US} to S'_{US} (Greenwood 1994).

The situation described above is one effect immigration can have on the demand for workers. There are many reasons for a downward and outward shift of the labor-demand curve from D_{US} to D'_{US} and D''_{US} (Greenwood and Hunt 1984). For example, the immigrants may be better educated and more innovative than the native population, and as a result they may alter the labor-demand curve. In some cases the immigrants may bring physical and financial capital with them, thereby altering the local economic structure and influencing the shift of the labor-demand curve. Immigrants also cause an increase in investments in infrastructure and housing, thereby shifting the labor-demand

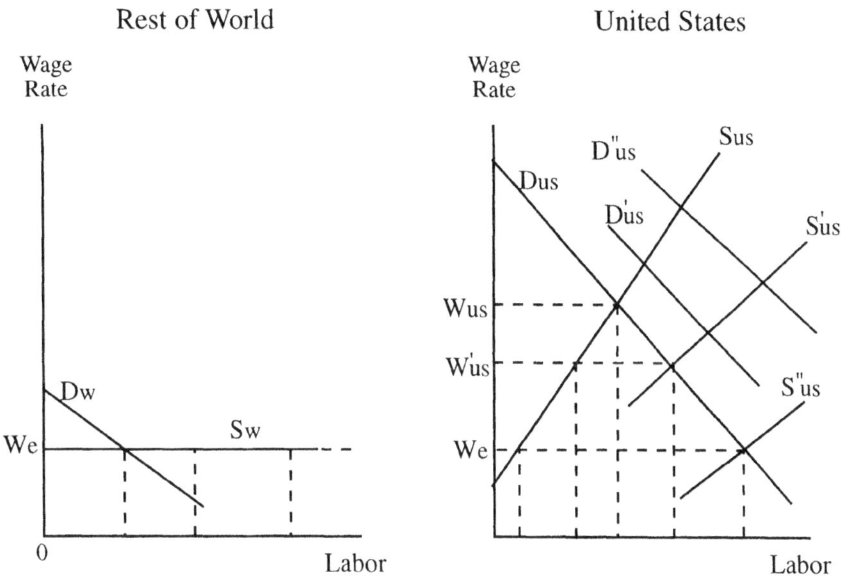

Fig. 3: Effect of immigration on wage rates (from Greenwood 1994)

curve. The shift of the demand curve from D_{US} to D'_{US} and D''_{US} due to these factors will *boost wages* above W_{US}. Theoretically, immigration to the United States will continue so long as there remains a disparity in real wages between the US and other countries. If immigration does not alter the demand for workers in the United States, and D_{US} remains the same, migration will ultimately boost the demand for labor (S''_{US}) and will bring real wages in the US to W_e, which is the global wage. In this theoretical case, a worldwide equilibrium will have been achieved for both wages and migration.

With respect to the salary level of immigrants, Poston's study on the financial achievements (income) of immigrant men in the United States is interesting (Poston 1994). Poston compares the financial achievements of men born in 92 countries other than the United States with seven categories of native-born Americans, such as Anglo-Saxons, African-Americans, and Asians. The financial achievements (chiefly income per worker) were examined against 14 independent variables, including country of origin, level of education, marital status, year of immigration to the United States, and proficiency in English. The results of the study bolster the thesis that the United States is a good melting pot for immigrants, as they assimilate fairly rapidly in the job market and sometimes earn more than the native-born. For example, the study found that 41 of the 92 groups of immigrants earn more than Anglo-Saxons and Asians born in the United States. There are also differences between groups of immigrants. European immigrants do much better financially than immigrants from elsewhere. Asians from countries that traditionally send immigrants to the United States, such as China, Japan, and the Philippines, do better financially than immigrants from other Asian countries. Another study has similar findings: European immigrants in the United States earn slightly more than the US average, and Asians earn slightly less than average. The average income of Latin American and African immigrants is substantially lower than the US average (Portes and Rumbaut 1990).

Massey sums up the main traits that affect immigrants' wages in the short, medium, and long terms (Massey 1987). Among them are the following:

- Level of education: The more education one has, the better one's chances are of earning a high salary in the medium and long terms.
- Age: The probability of obtaining a good salary is greatest between the ages of 25 and 44. Younger and older people are less likely to obtain high salaries.
- Work experience: Experience in the destination country is more important than in the country of origin. The more experience one has in the destination country, the better one's chances are of earning a high salary.
- Social and family ties: The presence of large communities from the country of origin and family ties help immigrants find jobs, "advance" at work, and receive fair compensation for their work.
- Length of time in the country: In the short term, immigrants get relatively low-paying jobs, but the longer they have been in the country, the higher their salaries rise, because they have had a chance to adapt their occupational skills to the supply of jobs.
- Knowledge of the language of the destination country: Knowledge of English is an important factor in integration in the US job market. Immigrants' ability to use their education and skills to best advantage in the job market depends partly on proficiency in the English language (Wallace 1986). In a comprehensive study of immigrants who moved to the United States in the 1960s and 1970s, Chiswick proves the importance of knowing the English language as a variable that explains immigrants' success in breaking into the job market (Chiswick 1984).
- Legal status: Illegal aliens earn less than legal immigrants because employers take advantage of their status and pay them lower wages. In the United States, illegal aliens do not view their stay in the country as permanent, so they do not invest in acquiring human-capital traits for which they would be remunerated in the labor market. Illegal aliens have relatively little motivation to acquire specific job skills for the American labor market, because, if they are deported or when they leave of their own free will, they may not be able to take advantage of these skills in their countries of origin. The illegal aliens think in the short term, unlike legal immigrants, who regard their stay in the United States as long-term, invest in acquiring job skills, and therefore have greater access to jobs that pay average to above-average wages (Chiswick 1984, 1986).

With respect to the wages of immigrants in the destination country, Borjas notes that when immigrants arrive in the destination country they lack traits and skills that are necessary in their new country's job market. Therefore their wages in the short term are lower than those of the natives. Over time, they invest in developing their human capital—at times more than the native population does. They do so by learning the language, acquiring additional education, and sometimes migrating within the destination country. This investment bears fruit over time and their income rises accordingly, sometimes more quickly than that of the native population (Borjas 1987a, 1987b, 1989). Borjas notes that the investment in human capital depends on the length of time immigrants intend to stay in the destination country. The longer they intend to remain, the more of an investment they will make, and the more rapidly their income will increase.

In the short term, Borjas argues, immigrants' wages are low. The increased supply of labor depresses marginal productivity of immigrants and thus wages per worker. It can also be argued that the decline in wages is a result of competition among immigrants in the job market. The graphic expression of this is a decline in the salary level of the immigrants themselves (from W_i^* to W_i') due to the supply of immigrants in the job market and the stability of the graph of demand for immigrant workers, especially in the short term (see fig. 4).

The effect of the immigrants on natives' wages—as was explained above in this section—depends on whether they substitute for or complement the native labor force. Which is the case? The conclusion must be reached through empirical research. If the immigrant labor force can substitute for the native labor force, then demand for local labor will fall. The demand curve will shift downward from D_i to D_i', and wages will fall from W_i to W_i'. If the immigrant labor force complements the native labor force in the production process, then the demand curve will shift upward from D_i to D_i'', and wages will rise from W_i to W_i'' (see fig. 4).

A number of empirical studies have been carried out in the United States on the impact of immigration on natives' wages (Grossman 1982; Borjas 1987b; Bean, Lowell, and Taylor 1988). The studies are based on different statistical sources and use different research methods. All reach the same conclusion: immigration has very little effect on the wages and job opportunities of natives, but a significant effect on the wages of the immigrants themselves.

The numerous hypotheses about the effect of immigration on the job market in the destination country are situated between two extremes: One extreme maintains that

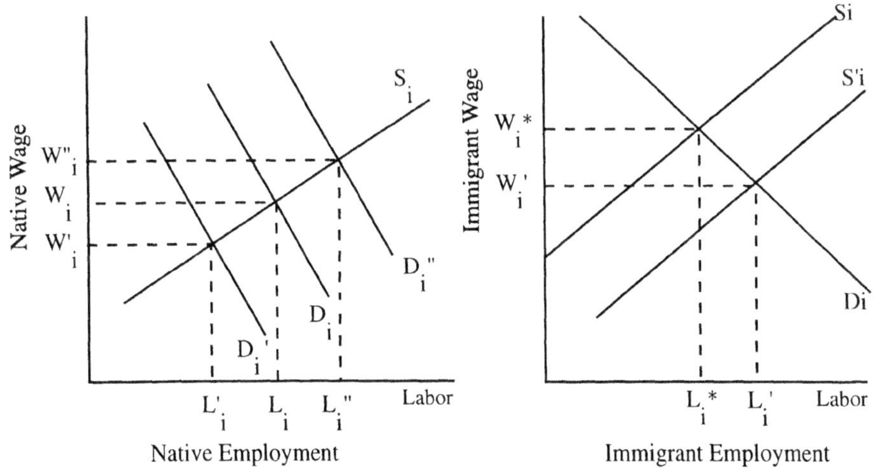

Fig. 4: Wages and employment of natives and immigrants (based on Borjas 1989)

immigrants take jobs away from native workers, in such a way that every immigrant forces a native worker to change his or her place of employment. The other extreme says that immigrants take jobs that natives/locals refuse (Greenwood and McDowell 1986; Borjas and Tienda 1987; King 1993c). According to Borjas, these two hypotheses should be examined in the empirical context in each country by asking the following question: are production factors—the native and immigrant work forces—substitutable or complementary? In the short term, the immigrants are probably not able to replace the native-born, because they do not know the language and are likely to have no financial capital and less education than the natives. In the long term, however, after they have acquired an education and amassed capital, they can certainly substitute for the natives. When the immigrants' level of education is similar to or even higher than that of the nonimmigrants—as, for example, in Canada in the 1980s and in Israel in the 1990s—the immigrants constitute a competitive substitute for natives even in the short term, and the greater the substitution is, the more negative the effect will be on wages (depressing them or preventing more than a slow increase). An examination of the degree of substitution and complementarity of the two production factors—the immigrant work force and the native work force—in the US labor market shows that there is some substitution, but not very much (Grossman 1982; Borjas 1983; Bean, Lowell, and Taylor 1988).

Beenstock and Fisher, citing other studies, argue that the theoretical literature provides no unequivocal answer to the question of whether the immigrants have a positive or negative impact on the employment situation of natives. Moreover, we certainly cannot conclude that the immigrants have a particular impact on *all* natives. Quite possibly, some immigrants act as substitutes and others as complements. The result depends on the immigrants' country of origin, their socioeconomic traits, and even the reason for their migration and the length of time they intend to remain in the destination country. It seems that only in unskilled work can it be assumed that immigrants are perfect substitutes for natives (Beenstock and Fisher 1996).

Piore (1979) argues that the immigrants take jobs that natives refuse to take. According to this approach, the job market can be divided into two sectors: "good" jobs and "bad" jobs. The natives choose the good jobs, and the immigrants are pushed into the bad jobs—which were rejected by the natives—due to their lack of skill or experience, language problems, and discrimination. If the immigrants and natives are complementary, then the natives will be freed to take jobs that require human capital, as they have a comparative advantage in such work. In this case, an increase in immigration will boost the natives' productivity, increase their wages, and improve their job situations (Friedberg and Hunt 1995). Although studies indicate that the effect of immigration on the American economy is marginal and even negligible, this appears to be because the economy is huge. Therefore, an addition to the labor supply as a result of immigration is not significant. Immigrants to Israel in the past five years (1990–1995) alone constitute a large proportion (about 15 percent) of the country's Jewish population, and therefore it can be assumed that the effect will be more significant.

Most researchers support the hypothesis that immigrants and natives are substitutable. In the short term the immigrants compete with native unskilled laborers; in the long term, after having acquired an education, learned the language, and taken vocational training courses, they compete with a different segment of the population, including skilled workers and academic and scientific professionals.

The effect of immigration on the economy of the destination country or of a particular region in the destination country is also manifested in technological changes produced by the immigrants. At the turn of the century, immigrants were the great entrepreneurs and innovators in the US economy. Another way in which immigration has brought about technological innovation—in the past more than the present—was via rapid population growth, which led to a need for increased production (a theoretical and empirical explanation of this can be found in the works of the renowned economist Kuznets; see, for example, Kuznets 1960).

Another way that immigrants integrate into the job market is by opening *small businesses*. Those who take such initiatives do so after the first stage of economic integration (Min 1984; Kim 1987; Portes and Rumbaut 1990). Most immigrant businesses (at least in the short and medium terms) require relatively little skill and investment and generally serve the immigrant community only. The owners hire family members or members of the same ethnic group, who are an inexpensive, flexible work force. In some cases the immigrants' entrepreneurial activity reaches beyond the community. For example, the Koreans in the United States do not limit their businesses to the Korean community; the Hispanics, on the other hand, tend to remain within the community (Salt 1992a).

Another way of integrating economically is through downward adjustment of one's occupation. For example, Korean doctors in the United States are willing to work in other health professions, and even in lower-prestige occupations in non-health fields (Shin and Chang 1988). By adapting in this way, they escape the ranks of the unemployed, become part of the productive population, and increase their income.

As immigrants integrate in the local economy, their unemployment rate drops. Immigrants have a higher unemployment rate than the native-born only during the first few years. In the United States and Australia, for example, the gap between the unemployment rates of immigrants and natives closes after ten years (Simon 1992).

One well-known phenomenon in international migration is the arrival of foreigners to work in the destination country. A prominent example of this is the migration of Turks to Germany (Kemper 1993, 1995). Until 1961 migration from Turkey to Germany proceeded at a slow pace. In that year—when there were about 4,000 Turks in West Germany—the two countries signed a labor-recruitment accord beneficial to both. West Germany, which was at the peak of economic prosperity, needed working hands, inter alia because the closure of the border with East Germany had halted the flow of migrants to the western side. In contrast, Turkey, following the assumption of power by a military regime in 1960, was contending with severe economic problems and worsening unemployment. Turkey was happy to go along with any idea that could help solve its labor problem. In 1973 approximately 2.6 million guest workers were employed in Germany; together with their families they numbered 3.9 million. In 1982 there were 4.6 million guest workers and their family members in Germany (Kemper 1993). These people were not and still are not eligible for German citizenship, and Germany now must confront the citizenship question, partly because of the rise of nationalist groups in Germany that perpetrate acts of violence against immigrants.

One of the major issues addressed in the literature on the impact of immigration on the economy is its effect on the public purse. Usher suggests estimating "net real wage," which is the actual wage less taxes paid by the immigrants, plus all the benefits received by the immigrants from the government. The net real wage—and not the actual wage earned by the immigrants—is the proper parameter for assessing the cost versus the benefits of immigration (Usher 1977). If the immigrants are relatively poor, a progressive tax will have a negative impact on the profits of the native population because the natives' contribution to supporting the immigrants (via taxes) will exceed the immigrants' financial contribution to the country. The composition of the immigrant population also affects natives' well-being. For example, if the immigrants have a relatively large number of school-age children, public expenditure will increase and the native population will be more heavily taxed, and the effect on the well-being of the native population will be negative (Greenwood and McDowell 1986).

Beenstock and Fisher (1986) note that there is a belief—similar to the belief that immigration is a burden on natives in the job market of the destination country—that the immigrants are a burden on the state. After all, their absorption entails increased government spending on education, health services, and welfare, whereas they contribute little to the public purse—especially in the short term—because they earn low wages and therefore pay low taxes. A study by Simon (1992) refutes this belief, or at least argues that it should be given little weight. His main point is that the immigrants' use of government welfare services is not a burden on the natives. Immigrant families use these services disproportionately little in relation to their share of the population. Using the Canadian example, Simon demonstrates that in their first three years in the country, demand by immigrants for financial support from the government decreases drastically, the proportion of immigrants below the poverty line falls dramatically, and the amount of income tax they pay increases.

The effect of immigration on wages can be spatially differentiated, too. If immigrants and natives are perfect substitutes for each other, the geographical distribution of the immigrants will depress wages in the destination regions or cause wages to rise there more slowly than elsewhere. If immigrants and natives complement each other in terms of production processes, then under certain conditions of elasticity of the labor-demand and -supply curves (i.e., when both are elastic), immigration will hike the wages of the natives and the immigrants and give a boost to economic activity.

Relatively little scholarly attention has been paid to the effect of immigration on the *economy of regions within a country*. Sometimes conclusions can be reached indirectly about the link between immigration and changes in the regional economy. One example of this is in Canada. Changes over time in immigrants' choice of destination in Canada have caused a demographic polarization between and within provinces, intensifying the geo-economic polarization (Lipshitz 1996a). The reason is not only accelerated population growth in a few central provinces due to immigration (which, of course, has an economic impact by raising the threshold and boosting purchasing power) but also—and perhaps primarily—the quality of the immigrants. In 1986 about 12 percent of immigrants to Canada aged 15 and over had university degrees, compared to 9 percent of nonimmigrants. About 16 percent of immigrant men are skilled workers, compared to 12 percent of nonimmigrant men. The longer immigrants have been in Canada, the higher their salaries are, and the greater their labor-force participation rate is. Among some age cohorts it even slightly surpasses the labor-force participation rate of the native population (Meckie and Thompson 1990). In other words, because they are educated, have technological professions, and are highly motivated to integrate economically, the immigrants are an important motivating force in the Canadian economy, and especially in the economies of certain provinces and cities. Because the immigrants are concentrated in a few regions, they provide impetus to the local economy in these places, while not affecting most other regions. In this way immigration increases interregional and intraregional disparities in Canada. It should be noted that the immigrants respond quickly to changes in the Canadian national and spatial economy. Although in the short term they tend to stay where they initially settled, in the medium term they migrate, especially to those few regions that are experiencing economic growth (Moore, Ray, and Rosenberg 1990).

1.6 Immigration Policy

Immigration policy—an issue of particular concern to politicians—is a governmental effort to regulate and control the entry of immigrants into the country, to set annual immigration quotas, and to establish the terms under which people can settle there permanently, work there temporarily, or obtain political asylum (Cruz 1991).

One of the fundamental questions in immigration policy issues is why such a policy is needed at all. Why can't everyone in the world live wherever he or she wishes, as was the case in most European countries before World War I (Fielding 1993b)? Such free migration is favored by economists from the neo-liberal school, which advocates free spatial movement of capital and labor and therefore opposes government intervention in international migration (Simon 1992). Other researchers who oppose government intervention in migration do so because of their conception of the individual's right to freedom of movement. Prevention of migration, especially from the Third World to the West, perpetuates suffering in the Third World and increases economic disparities between countries. As Zolberg writes, "In a world characterized by widely varying conditions, international borders serve to maintain global inequality" (Zolberg 1989, p. 406)

Another group of politicians and academics justifies the existence of immigration policy because without it it would be difficult to develop stable societal frameworks, and because of a fear of gradual invasion (Widgren 1990).

With the collapse of the USSR and the liberation of Eastern European countries from Soviet domination, Western Europe became an extremely popular destination for immigrants. Western countries that had always protested against the closed-border policy of the USSR were forced to confront the possibility of a flood of emigrants from these countries, in addition to refugees from Yugoslavia, which was disintegrating; meanwhile, Western Europe was suffering from an economic slump that elicited domestic opposition to immigration.

The changes in the European economy, especially the geopolitical changes, prompted each Western European country to impose its own immigration policy, reflecting its geographical location, its socioeconomic situation, and its political nature (Fielding 1993b). West Germany, for example, has encouraged immigration, especially by "guest workers." By 1973, 2.6 million immigrants of this sort had arrived in Germany. This number reflects the relations between Germany as a destination country and the immigrants' countries of origin (especially Turkey), as well as the growth of and structural changes in the German economy. The German rotation policy, under which immigrants were to live there for short periods of time, acquire job skills, and then return to their countries of origin, where they were to help develop those countries' economies, was more theoretical than practical. In fact, the immigrants found ways to stay (Böhning 1984). As of 1982, about half of all temporary immigrants in Germany had been there for over ten years, and all efforts to return them had failed.

In France immigration policy is unrestricted for residents of the former French colonies. Furthermore, France's attitude to the entry of immigrants (especially from countries in the Mediterranean Basin) was liberal for two reasons: demographic (very low natural increase in France) and cultural (dissemination of French culture). This policy changed after World War II, and France is now instituting a strict immigration policy (Hollifield and Ross 1991). One of the main demands of Le Pen's National Front in France is a halt to large-scale immigration to France and the expulsion of illegal aliens. Underlying this argument is the idea that the "immigrants" take jobs away from the "French."

Until 1962 all citizens of Commonwealth countries could immigrate freely to Britain. Since then quotas have been imposed to keep down the number of immigrants from Asia and the Caribbean islands.

The problem at present with regard to international migration is thus not the ability to emigrate (since the dissolution of the Communist bloc, very few countries prohibit emigration), but rather permission to immigrate. Improvements in means of transportation and their availability has cut travel costs and consequently increased movement of individuals, including migration. Worldwide economic development has produced cities with cosmopolitan, diverse populations, thereby creating a human infrastructure for immigrant absorption. Moreover, employers in destination countries favor immigration, which supplies them with an unlimited labor force at attractive prices. On the other side are the governments, which, partly under domestic pressure, oppose uncontrolled immigration because they want to preserve the country's cultural and historical hue, and because the immigrants take funds out of the public purse without ever having contributed to it through taxes. The conflict in industrialized countries is between the desire to maximize personnel and its flexibility and the desire to preserve the national culture and heritage (Fielding 1993b). Widening gaps in standards of living between Western Europe and the Third World have produced two opposite results: on the one hand the destination countries have tightened restrictions on immigration, and on the other hand the number of illegal immigrants has increased, as has the number of people seeking political asylum as the only possible way of "migrating" legally. In a study on immigration policy in several countries, Freeman (1992) found that the 1990s are characterized by "a strong pattern of restrictionism," in view of the mounting pressure of immigration applications.

In Europe policy measures are formulated from time to time, chiefly to stem the tide of illegal immigration and the movement of guest workers. Public opinion in destination countries—especially in Europe—is becoming more and more opposed to the presence of immigrants, especially non-European ones. The ethnic factor is crucial in Western European society. Demands that immigration be reduced or halted are made on ethnic grounds, and ethnicity is a factor in the renewed rise of national identity (White 1993). Various European countries, each in its own way, are trying to cope with refugees, political-asylum seekers, and guest workers, but with little success. In 1989, for example, Sweden—one of the most liberal countries—enacted a law that restricts the admission of guest workers and asylum seekers. It was repealed in 1991 due to criticism and pressure from minorities.

Since the 1970s European migration policies have been based on the premise that immigration can be controlled. After the recession of the 1970s and the oil crisis that followed, all Western European countries declared an "immigration stop" policy—a meticulous filtering of immigrants and granting of entry permits only to specific categories of people with needed occupations. However, this policy was not successful in most of the countries that tried it. Immigration to Western European countries actually increased during this period (Brochmann 1993). The immigrants found ways of circumventing the entry restrictions. As Brochmann writes (p. 108), "Control of one gateway may consequently direct the flow on to a new track."

Another factor that makes control of immigration to Europe difficult is the Maastricht Treaty, which turns Europe into a free-trade zone with no passport control at the borders. This makes it difficult to stem the flow of immigrants from one country to another and complicates the stringent policy measures that various countries want to use with regard to immigration.

Countries with long traditions of immigration—such as the United States, Australia, and Canada—have recently instituted more stringent policies regarding the admission of immigrants. The main concern of the US immigration policy is countering the massive amount of illegal immigration, chiefly from Mexico (Jones 1995). In this it has failed. NAFTA may reduce immigration pressures by increasing trade and eliminating the need to immigrate, provided that trade and immigration are substitutable (conversely, if they are complementary, then increased trade will boost immigration; see Razin and Sadka 1995). Some sectors in the United States favor immigration; among them are farmers and owners of light industry, who need people with little skill to work for minimum wages. The theory that competition in international trade is related to immigration is gaining strength in the United States. The theory maintains that international competition for production and trade increases the desire to take in immigrants as an efficient, production-boosting work force; because wages are low, low prices can be set for products, thereby increasing the probability of winning the competition in the global marketplace.

Like the United States, Australia had a pro-immigration policy for many years. Immigrants attained rights quickly, and the country encouraged their naturalization. The economic rationale behind the encouragement of immigration stemmed from the potential of a country rich in natural resources but short on labor. Moreover, strategic considerations of the "populate or perish" variety became important after World War II proved that the country was vulnerable to invasion from the north (Jones 1994). The general consensus that used to exist in Australia with respect to immigration policy is being undermined. Recently the Australian government adopted a series of measures to slow the pace of immigration and tighten control of it. The number of immigrants has indeed declined: from 140,000 in 1989–1990 to about 80,000 in 1992–1993. Interestingly, immigration to Australia increased in the 1970s when there was a recession there. The possible reasons for this are mainly political: pressure from the growth lobby of contractors who own labor-intensive businesses; economists' belief in the favorable effect of immigration on the economy; pressure by ethnic groups; and pressure by liberal intellectual groups that support the admission of immigrants, especially refugees and seekers of political asylum, on humanitarian grounds.

Canada, like Australia, recently considered making its immigration policy more stringent, primarily for economic reasons. Canada's immigration policy is more lenient with regard to skilled workers and academics. In the end, Canada did not curtail the number of immigrants. The official immigration plan set forth in 1990 for 1991–1995 encourages immigration by people with occupations that the country lacks and calls for directing some immigrants to areas where their occupations are needed. The Canadian policy thus falls somewhere between the US policy and the Australian one. It does not reduce the number of immigrants as the Australian one does, but it is more careful than the US policy is to link the immigrants' occupation and level of education to market needs and economic development.

Immigration policies encompass more than just immigration quotas. There are also social, economic, housing, and spatial policies. Few countries have a *spatial immigration policy* for the geographic dispersal of immigrants within the country, as Israel does (Shachar 1971; Lipshitz 1990, 1991a). Most countries leave geographic dispersal and employment to market forces. As for *sociocultural policies* on immigrant absorption there are two approaches: One is the "melting pot" approach, under which immigrants from all countries undergo an identical process of assimilation in the destination country. This conception is based on the argument that social assimilation is important for creating a new social basis for the country, and that this assimilation helps with economic advancement. The second approach is multiculturalism. This policy encourages retention of immigrants' ethnocultural identity and creation of a multicultural nation. The first approach is more dominant in the US, whereas the second is adopted more by Canada and Australia.

In Australia the first strategy was used until the 1960s; since the early 1970s the second has been dominant (Ozdowski 1985; Patterson 1990). In 1947 the Australian population numbered 7.6 million, only 9.8 percent of them foreign-born (and most of these were from the British Empire). Therefore the Anglo-Saxon culture was dominant and the language was uniform. Within 30 years the population doubled and its socio-ethnic composition changed radically. In 1984 Australia had a population of 15.6 million, 3.2 million of them (21 percent) born in foreign countries, among them Eastern European, Middle Eastern, and South American states (Patterson 1990). These waves of immigration altered the sociocultural structure in Australia. One manifestation of this change is the multiplicity and diversity of the media. Today there are close to 80 ethnic newspapers and periodicals in 25 languages, in addition to a wide, multicultural range of radio and television broadcasts (Patterson 1990). The radical changes in the ethnic makeup of immigration to Australia forced policymakers to switch from a melting-pot policy to multiculturalism.

There are two main approaches to *housing policy*. Some countries do not intervene at all in the housing of immigrants; they leave the work of housing to the free market. Some Western European countries intervened in the immigrant housing market in the past, building neighborhoods for immigrants. This approach increased sociocultural segregation within cities. In several Western European countries—especially West Germany after World War II—immigrants were housed in suburban neighborhoods near large factories. The homes were provided by the employers (King 1993a; White 1993). Recently several Western European countries and cities have taken steps to prevent immigrants from settling in certain neighborhoods and housing projects in order to eliminate segregation, but without success (van Amersfoort and de Klerk 1987; Sporton and White 1989; Arin 1991; White 1993). Since the start of the 1970s and even more so in the 1980s, some public housing has been privatized, and less public housing has been built for immigrants (Krätke 1989). The free housing market in Western Europe (e.g., the Netherlands) reduced housing opportunities for immigrants, especially illegal immigrants and refugees of the third (post-industrial) wave. As a result, in some European countries a subsector is emerging in the housing market, based on organization by ethnic community, in an effort to help immigrants from the same ethnic group find suitable housing (White 1993).

To sum up, this chapter looked at trends of international migration with regard to the following topics: changes in the geographical patterns of international migration; the geographical distribution of immigrants within countries; internal migration by natives (and immigrants) as a result of immigration and as a result of economic and social factors; the economic impact of immigration on the destination countries; and official immigration policies. This survey serves as the background for the rest of the book, which will consider the spatial processes of immigration and internal migration in Israel. In the conclusion to the book we will take a look, inter alia, at the similarities and differences between migration trends in Israel and elsewhere.

CHAPTER 2

IMMIGRATION TO ISRAEL IN THE 1950s: CORE VERSUS PERIPHERY

2.1 Pre-State Waves of Immigration and the Geographical Distribution of Settlements

In order to understand the geographical distribution of the immigrants of the 1950s and the government policy that produced it, one should be familiar with the geographical distribution of the population on the eve of Israeli independence in 1948 (for a thorough discussion of this subject, see Kellerman 1993). This distribution was shaped by pre-State waves of immigration that began in 1882. The year 1882 marked the renewal of Jewish settlement in Palestine. Until then there were organized Jewish communities in only four cities in Palestine: Jerusalem, Tiberias, Safed, and Hebron.

Between 1882 and 1947, approximately 455,000 people moved to Israel in five waves of immigration. These waves (1882–1948) and their geographical distribution in Palestine are described here in brief.

1882–1903: The *First Aliya*, in which about 25,000 Jews moved to Palestine. This wave grew out of the first "Zionist" movement, Hovevei Zion, in Russia and Romania, and was sparked by a series of pogroms in southern Russia in 1881. The immigrants regarded settlement on the land as the paramount condition for the rejuvenation of the Jewish people (Eisenstadt 1973). They established Jewish agricultural settlements in Palestine, including the *moshavot* of Petah Tiqva, Rishon Lezion, Zikhron Ya'aqov, and Hadera.[1]

1904–1914: The *Second Aliya*, in which 40,000 Jews moved to Palestine. The immigrants of the Second Aliya were mainly "labor Zionists" from Russia who were disillusioned with the Russian social reform movements, in which they had taken an active part, and by the October 1905 revolution, which ended with pogroms against the Jews. It is often considered to be a workers' immigration movement. During this period new forms of settlement emerged (the moshav and the kibbutz)[1] and the foundation was laid for the Palestine labor movement.

1919–1923: The *Third Aliya*, in which 35,000 Jews moved to Palestine. This wave of immigration began after the Balfour Declaration (written by British Foreign

1. A *moshava* (pl. *moshavot*) is a rural settlement in which land is privately owned by the farmers. Production and consumption are private and subject to private decisions. The *moshav* (pl. *moshavim*) and *kibbutz* (pl. *kibbutzim*) are collective forms of rural settlement in which land is owned by the Jewish National Fund. On a moshav, all farmers have plots of land of equal size. Production on a moshav is each farmer's own responsibility (although the farmers have a coordinating body), members have land adjacent to their homes, and consumption is private. On a kibbutz, in contrast, production and consumption are collective (although recently significant changes have been made in this regard). One of the most important principles of the kibbutz is that each member produces according to his or her ability and consumes according to his or her needs. Kibbutz land is not divided among the members; instead, land is held communally and is scattered in very large plots around the kibbutz.

Secretary Arthur James Balfour), which the Jewish world interpreted as a new beginning in the realization of the Zionist ideal. Many of the newcomers were young people who were willing to do any work required of them. Large numbers of them turned to the moshavim and kibbutzim. Many of them were well-educated intellectuals.

1924–1931: The *Fourth Aliya*, in which 85,000 Jews moved to Palestine. There were two main motivations: the economic situation in Palestine had improved, making the absorption of additional immigrants feasible; and the economic situation of Polish Jewry had worsened due to the Polish government's policy of dispossessing Jews from major sectors of the economy. The immigrants were primarily from the middle class; most of them settled in the cities and found work in commerce, crafts, and industry.

1932–1947: The *Fifth Aliya*, in which 270,000 Jews moved to Palestine, mostly from Germany and Poland. These immigrants brought with them relatively large amounts of capital and developed industry, commerce, and agriculture in Palestine. This wave of immigration gave a great boost to the economy of Palestine and was a major milestone in the growth of the urban population, especially in the major cities (Tel Aviv and Haifa) and their periphery. Many of the immigrants who came at this time, especially in 1932–1935, had seen the writing on the wall with the accession of the Nazis to power in Germany, and decided to emigrate from Germany and eastern Europe.

The pre-State waves of immigration substantially increased the number of rural and urban settlements and the population. Between 1922 and 1945, the number of Jewish towns rose from 8 to 27 and the total population of these towns increased from 70,000 to 440,000; the number of moshavot rose from 34 to 44 and their population increased from 12,000 to 80,000; the number of moshavim rose from 11 to 98 and their population increased from 1,400 to 31,000; the number of kibbutzim rose from 19 to 115 and their population increased from 1,200 to 37,400 (Eisenstadt 1973).

Before 1948, every wave of immigration left its *geographical imprint* on the distribution of localities, the distribution of the population, and forms of settlement. In general, the first three waves left their imprint on agricultural settlement by founding moshavot, kibbutzim, and moshavim. In contrast, the Fourth Aliya was a milestone in the immigration of urban elements and the development of the urban sector, whereas the immigrants of the Fifth Aliya settled primarily in the urban centers and introduced new sectors and new kinds of economic enterprises (Eisenstadt 1973).

For ideological reasons and due to constraints posed by the need to purchase land from Arabs, a large proportion of the early collective agricultural settlements were established in geographical regions that would later become the *geographical periphery* of the State of Israel. *But this geographical periphery was by no means a socioeconomic periphery*. On the contrary, during the pre-State period, "ties between the elites and the agricultural settlements were very close. . . . Large portions of the political elite came from the agricultural settlements, and many national missions were carried out by the agricultural settlements" (Eisenstadt 1973, pp. 137–138). Statistics show that the sociocultural level of the pre-State agricultural settlers was quite high. In contrast, the settlement of immigrants in peripheral agricultural localities in the 1950s created a socioeconomic periphery. For example, in 1961 the median number of years of education was 9.9 on old moshavim (those founded before 1948), 10.9 on old kibbutzim, and only 6.8 on moshavim settled by immigrants in the 1950s.

The immigrants who chose urban localities in the pre-State period—and especially during the British Mandate (1917–1948)—usually settled in pre-existing cities in the central region. Two factors encouraged this tendency:

> First, [due to] economies of scale and the advantages of proximity . . . the immigrant entrepreneurs and immigrant workers preferred to concentrate in the coastal cities because these were centers of investment and job opportunities. Second, services in the major cities expanded as transportation conditions improved during the Mandate period. (Reichmann 1979, pp. 80–84)

The concentrated structure of the agricultural settlement movements also played a part in attracting immigrants to the central region: their main institutions, located in Tel Aviv, gave them countrywide purchasing, marketing, and financial services. As the quantity of services offered increased, additional jobs were created in the major cities, and this was a major factor in attracting immigrants.

The geographical distribution of pre-State immigration—especially the Fourth Aliya and the Fifth Aliya—led to accelerated urban development in an area that would later come to be called the Tel Aviv conurbation (on the development of this conurbation, see Amiran and Shachar 1971). Within a short period of time (1930–1937), the population of Tel Aviv grew by 108,000 (from 42,000 to 150,000), and the city began to spread northward and eastward. Amiran and Shachar state that this rapid growth was chiefly the result of the Fifth Aliya, which was decidedly urban in nature. The extensive development of the built-up area in the city of Tel Aviv during this period was another outcome of the Fifth Aliya and the economic means of its immigrants (Amiran and Shachar 1971). Concurrently, the urban localities around Tel Aviv grew, and the region began to look like a metropolitan area. The immigrants of the Fourth Aliya and Fifth Aliya, who were generally capitalistic, chose to live in the cities in the central region and in surrounding moshavot (see Gonen 1978). This process was one of the main causes of the initial urbanization of the moshavot (converting them from rural villages into cities). Because land in the moshavot was privately owned, the owners took advantage of the demand for housing and employment and permitted the construction of homes and workshops on this land. Soon "urban enterprises"—workshops, shops, and small businesses—began to take shape in these moshavot. The rapid transition of the moshavot from rural to urban was possible chiefly because they were "private" and not "national" rural settlements like the moshavim and kibbutzim. Urbanization occurred rapidly in the moshavot on the periphery of the Tel Aviv metropolitan area and along the Tel Aviv–Haifa axis, and extremely slowly (if at all) in the moshavot in the Galilee (northern Israel). The location of the former is responsible for the difference in urbanization processes. Proximity to Tel Aviv and to the Tel Aviv–Haifa "urbanization and development axis" attracted the immigrants of the 1930s to these moshavot; later, when suburbanization supplanted urbanization, these moshavot drew migrants from the city of Tel Aviv (Gonen 1978). In the 1960s and 1970s, as a result of these processes, many of them became medium-sized and large cities in the outer ring of the metropolitan area.

Thus, the Fourth Aliya and the Fifth Aliya caused a marked change in the geographical distribution of the Israeli population. Through their clear preference for the central region, they inaugurated *the economic and demographic core of Israel*. It can be said that the pre-State immigrants were absorbed through two mechanisms: planning from above (settlement institutions), which especially typified the Second Aliya and the Third Aliya, and planning from below (market forces), which was particularly characteristic of the Fourth Aliya and the Fifth Aliya. In terms of the geographical distribution of the immigrants, market forces had a more decisive impact. In other words, the dominant processes in the housing market were that immigrants purchased land on which to build their homes or purchased/rented homes. Ultimately, tenanting from above (by the Jewish Agency and other organizations) was relatively unimportant. As a result, when Israel gained its independence the population—most of it made up of pre-State immigrants—was highly concentrated in the central region.

Settlement by immigrants in the cities of the central region and the conversion of the moshavot in the central region into cities—a process that picked up speed due to immigration—polarized the spatial distribution of the country's population. The bulk of the Jewish population lived in a large number of cities in the central region, while the minority was scattered in a large number of agricultural settlements in the geographical periphery of the country, where such localities were the dominant form of settlement. Thus, when the country gained its independence, more than half of the population lived in Tel Aviv, one-fifth lived in Haifa, and one-tenth lived in Jerusalem. The share of the Northern District had dropped, and the Southern District was almost devoid of Jewish population (table 1 and figs. 5 and 6; see also Shachar 1971).

Table 1: The Geographical Distribution of Localities (1949) and Jewish Population (1950)

	Urban and rural localities		Urban		Rural		Urban + rural	
District	Number	Percent	Population	Percent	Population	Percent	Population	Percent
Jerusalem	19	3.9	120,000	13.71	3,232	3.78	128,232	11.74
Northern	172	35.1	33,565	3.84	61,190	28.10	94,755	8.68
Haifa	44	9.0	164,958	18.87	24,284	11.14	189,242	17.34
Central	193	39.4	97,954	11.21	102,923	47.25	200,877	18.40
Tel Aviv	24	4.9	449,491	51.42	11,749	5.41	461,240	42.21
Southern	38	7.7	8,300	0.95	9,430	4.32	17,730	1.63
Total	490	100.0	874,268	100.0	217,808	100.0	1,092,076	100.0

Based on a reworking of tables 4 and 5 from the *Statistical Abstract of Israel*, no. 2, 1951.
See the map of districts and their division into core and periphery in chapter 3 (p. 66).

Table 1 shows clearly that the geographical distribution of the urban population at the end of the 1940s (1950) was the main cause of the geographical polarization of the Israeli population. Although only 9 percent of Israeli localities (44 out of 490) were urban in 1949, the urban Jewish population accounted for 80 percent of the total Jewish population. As the table indicates, the urban population was concentrated in a small number of localities in the central region. For example, in 1949 about 5 percent of localities were in the Tel Aviv District, whereas 42 percent of the Jewish population lived in this district. The urban Jewish population in this district accounted for 51 percent

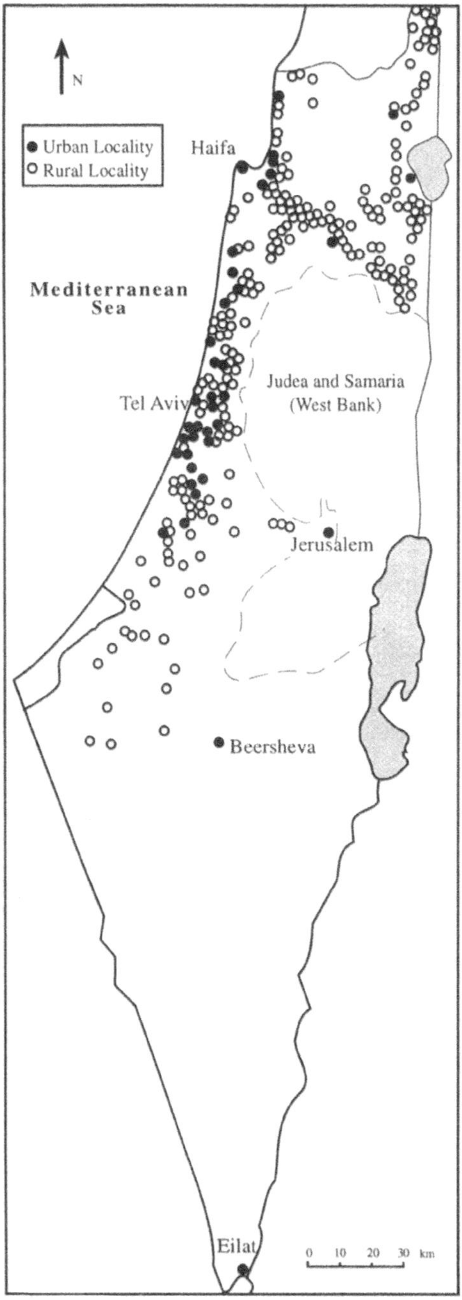

Fig. 5: The geographical distribution of urban and rural localities in 1948

of the total urban Jewish population in Israel, whereas the rural Jewish population in the Tel Aviv District accounted for 5.4 percent of the total rural Jewish population.

Interestingly, although in 1949 about 35 percent of Israeli localities were situated in the Northern District, only 8.5 percent of the Israeli Jewish population lived there. The main reason for this is the extremely high proportion of rural localities in the Northern District (over 95 percent of the localities in the district), as only a small number of people lived in each rural locality. In 1950 the rural Jewish population of the Northern District accounted for 28 percent of the total rural Jewish population of Israel, whereas the urban Jewish population of the Northern District accounted for 3.8 percent of the urban Jewish population. Overall, 8.6 percent of Israeli Jews lived in the Northern District (table 1). These examples are sufficient to illustrate that the geographical polarization of the Israeli population at the end of the 1940s was the result of the geographical polarization of the urban population, most of whom lived in a few localities in central Israel.

2.2 Spatial Policy for Dispersing the Immigrants of the 1950s

The polarized spatial distribution of the Jewish population, described in section 2.1, became an economic and geopolitical-security problem when the country gained its independence. Consequently, the Israeli government regarded a "population-dispersion policy" as one of its basic principles and as the main consideration in the geographic dispersion of the immigrants of the 1950s. The population-dispersion policy was made possible by the arrival of a million immigrants in the 1950s, for whom transit camps, new towns (known as "development towns"), and moshavim were built; most of the new towns and moshavim were built in the Galilee (in the north) and the Negev (in the south). *The Israeli population-dispersion policy of the 1950s was essentially an immigrant-dispersion policy.*

Such a large number of immigrants had never reached the country within such a short span of time in the pre-State period, nor has it since the 1950s. Even compared to other countries that admit large numbers of immigrants, it was a wave of impressive dimensions. The State of Israel, which at the time of independence had a Jewish population of close to 717,000, took in roughly one million immigrants in the space of ten years, including 724,000 in 1948–1953. The immigration of the 1950s transformed the socio-demographic structure of Israel and had an impact on its economic and geographical structure, as we shall see below (Sicron 1957).

The immigration of the 1950s, especially at first, was perceived by the country's leaders—Prime Minister David Ben-Gurion in particular—as an extremely important component of Israel's national security and military power (Segev 1984). Ben-Gurion made no secret of the fact that he considered the immigration crucial.

> We might have been able to capture the Triangle [an Arab-populated area in east-central Israel], the Golan, [and] the entire Gali-

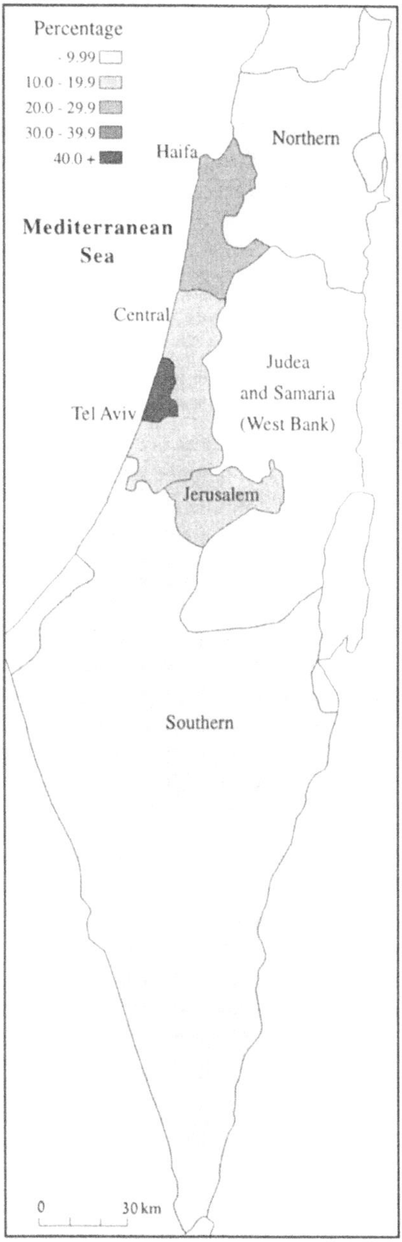

Fig. 6: The geographical distribution of the Israeli population in 1948
(population of each district as a percentage of the total Jewish population)

> lee, but conquering these would not have bolstered our security to the extent that the absorption of immigrants can. . . . Without settlement, conquests have no decisive value, not in the Negev, not in the Galilee, and not in Jerusalem. Settlements are the real conquest.

And because settlement depended on immigration, Ben-Gurion said, "The fate of the State depends on immigration" (Segev 1984, pp. 106–107).

The reasons for the population-dispersion policy in Israel were geopolitical (a Jewish presence in desolate and sparsely populated areas along the country's frontiers), security-related (the establishment of localities along the borders as part of the country's defense system), and economic (exploitation of mineral ores and other natural resources located in the periphery). Therefore, the policy was an important element of the guidelines of all Israeli governments. The guidelines of the first government said that the development programs of the next four years (1948–1952) would be geared toward doubling the Jewish population through mass immigration and intensive development based on a planned economy. The main goal of the development program was population dispersion and prevention of overconcentration in the major cities in central Israel. The intention was not only to establish a large number of localities all over the country, but to encourage settlement in agricultural localities in the periphery.

The population-dispersion program had five main elements (Sharon 1973): dispersion of agricultural settlements based on exploitation and development of water sources; a more balanced geographical distribution of industry and development of kinds of industry that did not have to be in the central region; the building of roads to desolate areas; establishment of parks and forests and development and preservation of the landscape; and establishment of new towns. The program proposed integrating the new urban and rural localities into the network of Israeli localities such that five types of localities would exist: the rural unit, with up to 100 agricultural families; the rural center, which would be a service center for several agricultural settlements; the town (an urban-rural center), with a population of 15,000–20,000 and services and factories for the residents of the town and the surrounding area; the medium-sized city, which would have a population of about 100,000 and would be the subdistrict "capital," with services and a high threshold for services and businesses; and the big city. The last-mentioned category refers to the three major Israeli cities, Jerusalem, Tel Aviv, and Haifa, which had an economic base and services on the national level.

Most of the immigrants who arrived in the 1950s came from North Africa and Asia. They arrived destitute, and many (about half) were uneducated; only a few (about 2 percent) had a higher education. The immigrants' economic situation and poor housing conditions in their countries of origin have been described in detail in reports by immigration workers who visited these countries. One of the reports, cited in Segev's book, says, "From hearing or reading about it, one would not believe that such conditions really exist in our world" (Segev 1984, p. 117). The result was that the immigrants were dependent on public (governmental) agencies for housing, employment, and such services as education and health care. Therefore, the government had almost exclusive control over the geographical placement of the immigrants, and for the reasons described above it chose the peripheral districts.

MEANS OF DISPERSING THE IMMIGRANTS

The government used a variety of means to disperse the immigrants throughout the periphery (Shachar 1971; Zilberger 1973):

Establishment of Transit Camps

The discrepancy between immigration forecasts and reality led to the establishment of immigrant camps, which in the late 1950s contained approximately 150,000 people, most of them living in tents (Hacohen 1994, p. 178). The immigrant camps were set up hurriedly in various parts of the country. Other immigrants were housed in Arab neighborhoods in the major cities that had been abandoned during the War of Independence.

Because of the poor physical conditions in the immigrant camps, a large-scale housing program was required. The immigrant camps were not suitable for protracted residence, and therefore it was decided to dismantle them and establish transit camps in their stead. The transit camps were intended to house the newcomers in improved conditions (wooden shacks instead of tents) until they obtained permanent housing. The idea was to free the immigrants from financial dependence on the Jewish Agency (an organization that served as a quasi-government in the pre-State period and subsequently has been involved, inter alia, in immigration and rural settlement) and the government. In the tent camps the immigrants had been passive with respect to employment and had been supported by the Jewish Agency; in the transit camps they were expected to work in the adjacent locality, while integrating into the social life of the country (Hacohen 1994).

The establishment of the transit camps—which provided temporary housing for hundreds of thousands of immigrants in the early years of the State—was one of the biggest public projects ever planned and carried out by the State of Israel to implement its sociospatial and housing policies (Felsenstein and Shachar 1986). Numerous transit camps were set up all over the country to resolve the severe housing crisis that had developed following the huge wave of immigration of 1948–1956. Immigration created a demand for housing on a tremendous scale that could not be met using the means and budget available to the Israeli government at the time.

The establishment of the transit camps—especially those in the Israeli periphery—was intended to achieve two goals: to solve the immigrant housing problem (albeit temporarily), and to implement in part the population-dispersion policy (Felsenstein and Shachar 1986). This idea was in sharp contrast to the situation in many countries, in which immigrants tend to settle in the big cities. The population-dispersion policy called for establishing some of the transit camps in the periphery in order to prevent the immigrants from concentrating in the major cities.

But plans and real life are two different things: in the end, most of the transit camps were set up in the central region (see fig. 7), some of them in former British army camps (from 1917–1948). Although jobs were in short supply and unemployment was high, the prospects of finding some sort of work in the future were actually quite good. Nevertheless, by joining the labor market in central Israel en masse, the immigrants caused wages in general to fall (Hacohen 1994). Thus, within the overall population-dispersion program opposing forces worked simultaneously to concentrate the

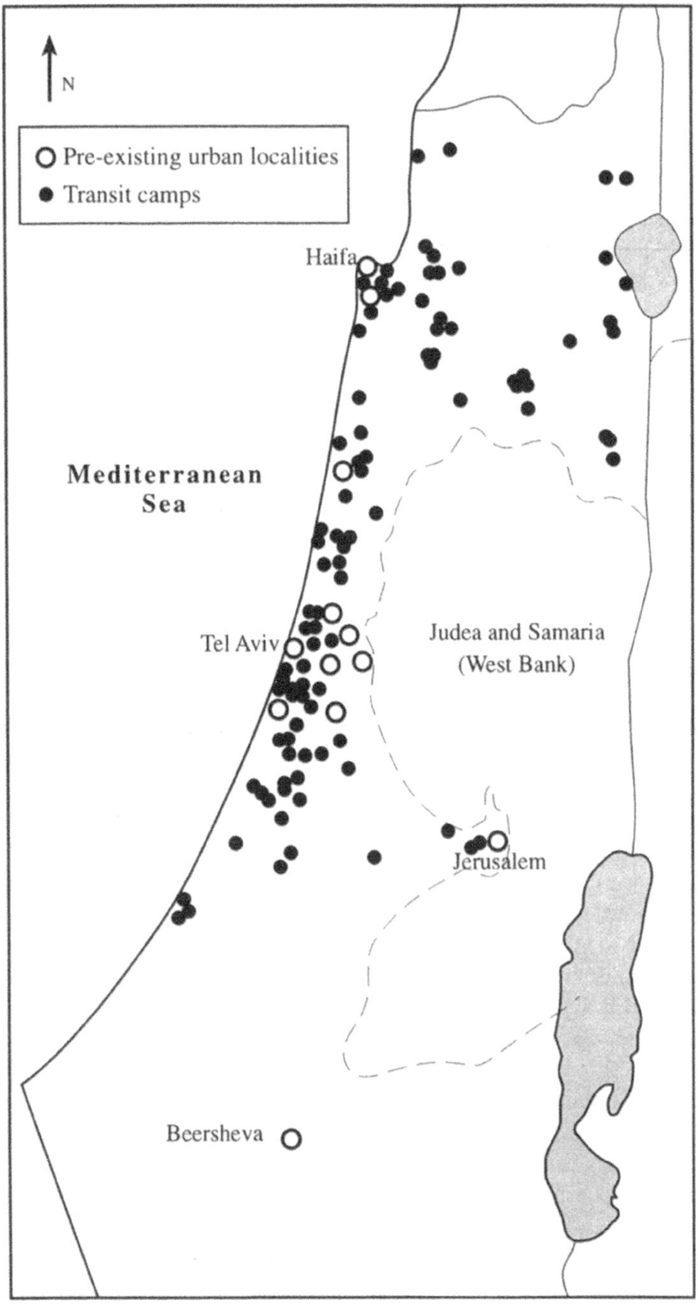

Fig. 7: The geographical distribution of transit camps established to house the immigrants of the early 1950s (based on Felsenstein and Shachar 1986)

immigrants (for reasons of employment and social integration) and to disperse them (for political and security reasons).

Some of the transit camps eventually evolved into towns. The camp at the village of Khalisa (in the north, on the Lebanese border) became Qiryat Shemona; the camp by Hartuv (between Tel Aviv and Jerusalem) became Beit Shemesh; and Or Yehuda (in the eastern part of the Tel Aviv metropolitan area) arose out of several transit camps. Other camps became poor neighborhoods on the fringes of medium-sized and large cities.

As of mid-1952, the transit camps and immigrant camps housed approximately 250,000 people, about one-sixth of the Jewish population of Israel (Hacohen 1994, p. 299). At the end of the year, the government decided to devise a plan for abolishing the transit camps and moving the immigrants to permanent housing in towns and agricultural localities. By 1954, the population of the transit camps and immigrant camps was down to 88,000. While the transit camps were being eliminated, there began a better-planned and more orderly process for housing the immigrants: the establishment of "development towns" (new towns) and "immigrant moshavim."

Establishment of New Towns

Thirty-two new towns, also known as "development towns," were founded in 1948–1962 (including a substantial expansion of existing towns, such as Safed, Tiberias, and Afula). In the early 1960s immigrants constituted 70 to 90 percent of the total population of many development towns. Most of them were really new, although a few—especially in the central region—were existing towns that were substantially enlarged during the 1950s (Shachar 1971). The geographical location of the new towns was intended to change fundamentally the unbalanced geographical distribution of 1948. Therefore, 12 towns were founded or greatly expanded in the north, 13 in the south, and 9 in the center (see fig. 8). By settling in these widely dispersed towns, the immigrants of the 1950s lessened the polarization of the geographical distribution of the population, as we shall see below.

The development towns were intended, inter alia, to eradicate the polarized distribution of localities in Israel, as they were somewhere between rural localities and the major cities. Thus they played a major role in converting the system of localities in Israel from a system of "primary cities" to a ranked system (Shachar 1971). This change, according to Shachar, is a unique case of planned policy working a rapid change in the urban ranking system (Shachar 1971).

In some parts of Israel development towns were also intended to serve as regional centers (service centers) for the rural population, providing services that are not available in rural localities because they require a relatively high threshold (Brutzkus 1973; Gradus and Stern 1980). Studies from the 1960s and 1970s show that where there was a fairly high concentration of moshavim (especially immigrant moshavim)—but not kibbutzim—around a development town, the development town usually served as a central location that provided services to its rural surroundings (Kipnis 1975). In the north, the economic ties between the development towns and their rural surroundings were weaker because of old kibbutzim in the area that preferred—apparently for ideological reasons—to get their services from Tel Aviv and Haifa rather

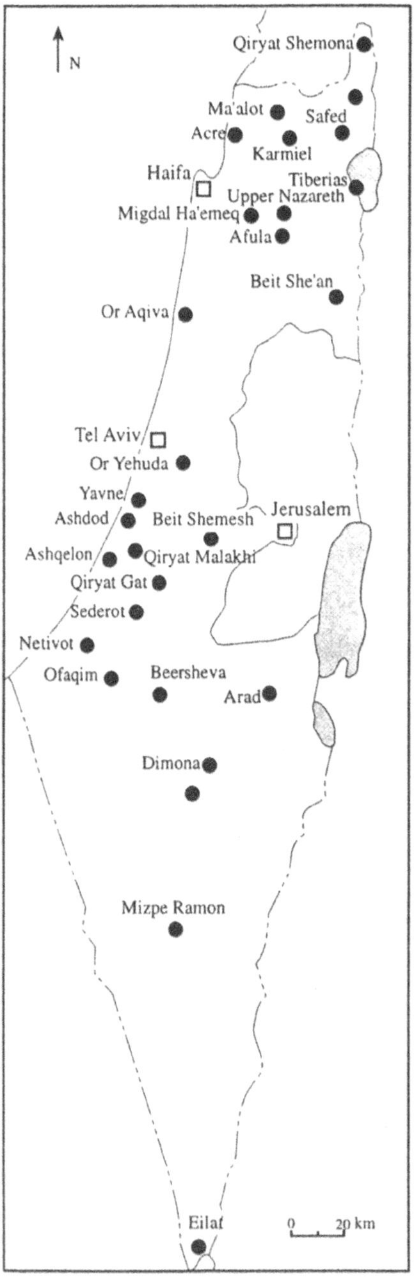

Fig. 8: The geographical distribution of Israeli towns founded for the immigrants of the 1950s
(including towns founded before 1948 that were enlarged greatly during the immigration wave of the 1950s)

than from the new towns (Kipnis 1983, 1984). The kibbutzim had a socialist world view, and the most extreme of them had previously refused to have anything to do with the capitalist city. Because of the economic structure of the kibbutz, purchases were handled by the kibbutz purchaser, who would go shopping several times a month. For economic reasons and reasons of convenience (a diverse supply of products) and efficiency, the purchasers would shop in Tel Aviv or Haifa.

In the development towns, the construction of small apartments for immigrants by the government preceded the creation of jobs. Most of the apartments measured between 36 and 50 square meters and conformed to a small number of models. The stage of job creation, which was based mainly on industrialization of the development towns, began later.

Government policy in the 1950s touched on industry as well as housing. This policy raised the "industrialization level" (the percentage of people employed in industry) of the development towns; nevertheless, industrial diversity in these towns was low, chiefly because of the tendency to open large, labor-intensive factories, producing food, textiles, leather, and leather goods (Gradus and Krakover 1976, 1977). The industrialization policy for the development towns solved the unemployment problem of many unskilled workers, but left another class of workers, albeit smaller but with a higher level of vocational skill, without suitable jobs (Bar-El 1982).

Two decades after the establishment of the development towns and their industrial system, the "traditional industries" were still dominant (Kipnis 1984, 1989). In 1976, 30 percent of industrial workers were employed in "advanced industries," and 70 percent worked in "traditional industries" (Borochov 1989). Chemical plants in the Negev, which are based on the quarries in the region, accounted for much of the employment in advanced industries in development towns. Without these plants, even in the mid-1970s a low proportion of development-town residents worked in advanced industries. It should be noted that one of the major features of the "traditional industries" is low wages. Moreover, these industries do not develop, and therefore they will not provide additional jobs in the future. Often, these factories (especially textile plants) suffer as a result of competing imports, and therefore employment runs the risk of declining.

Not only were many development towns characterized by a high percentage of people employed in industry; in addition, industry in the development towns had a concentrated structure, i.e., a large number of workers in a few factories. In many development towns over half the industrial workers were employed by factories with 500 or more workers; in some towns, 40 to 75 percent worked in three or four factories. This structure made employment in the development towns quite vulnerable, as closing a single factory was enough to raise unemployment rates considerably in a particular town in a relatively short period of time. This scenario occurred in several towns. Many of the immigrants from the 1950s, and even the second generation, often had serious fears that their factories would shut down. Sometimes the fears were substantiated, as large factories in the development towns closed and the workers were fired. In other cases political pressure resulted in government bailouts to prevent factories from closing.

Another substantial portion of the work force in development towns was employed by the public-service sector (e.g., local and central government, welfare services, education). In some development towns even more people were employed in the public services than in industry, and the percentage exceeded the national average. "Swelling" the public sector in development towns beyond what is needed is a relatively easy way to create jobs. The central and local governments often chose to develop the public sector more than the communities needed, so as to provide jobs for the immigrants of the 1950s and their children.

Thus, in the development towns that were founded primarily to absorb the immigrants of the 1950s, labor-intensive industry was established in accordance with the immigrants' occupational and educational profile. These industries and public services paid low wages and did not produce a multiplier effect with respect to employment and income. This had ramifications for population growth trends in these localities and for the socioeconomic gaps between cities in central Israel (the core region) and the development towns (the peripheral region), as we shall see in the next chapter.

The development towns of the 1950s and 1960s resembled each other not only in terms of housing and employment, but also in terms of ethnicity, society, and culture. The bulk of the population of these towns had immigrated from North Africa and Asia. In many development towns in the early 1960s, approximately 90 percent of the population had come from these countries.

Establishment of Agricultural Settlements

In the twentieth century, most immigrants worldwide have found work in crafts and industry in the big cities and their periphery. Israel chose instead to direct the immigrants to rural settlements, where they would work in agriculture. About 200 rural settlements, most of them moshavim, were founded to house the immigrants of the 1950s. There are three reasons for this (Weitz and Rokeach 1986): First, a Zionist tradition, more than 50 years old, associated the revival of the nation in its own independent state with its return to the land as workers of the land. Second, from a geopolitical-security perspective, it was necessary to settle extensive tracts of land that were uninhabited following the War of Independence. Third, the Arab economy, which had supplied the bulk of agricultural produce in Palestine during the Mandate period, had been severely impaired during the War of Independence, as had the Jewish agricultural sector. As a result, there was a severe shortage of all agricultural products, and the scarcity became even more problematic in the wake of the mass immigration, which doubled the Jewish population of Israel within just a few years. Therefore, extensive national resources were directed toward developing the agricultural infrastructure, and many new immigrants were placed in rural settlements and trained to work in agriculture. Approximately 200 new agricultural moshavim were founded between 1949 and 1951, and by 1954 they had about 60,000 inhabitants (Weitz and Rokeach 1986). A total of 282 moshavim were founded during the 1950s. Most were populated by new immigrants: about 90 by immigrants from Europe and 160 by immigrants from North Africa and Asia (Weitz and Rokeach 1986). Most of these moshavim were established in the periphery.

Among the agricultural settlements there were differences of opinion about immigrant absorption in the 1950s. The kibbutzim were concerned that taking in large numbers of immigrants would undermine their socialist and ideological foundations, whereas the moshav movement (whose socialist ideology was less extreme, to the extent that some of the moshavim were even moderately capitalistic) adopted a much more positive approach. The leaders of the moshav movement acceded to the government's request and agreed to accept immigrants. They were motivated by national and societal considerations, as well as the belief that they could work a cultural and social transformation among the immigrants, turning them into farmers and integrating them in the rural settlements (Hacohen 1994). As a result, the government concentrated on establishing moshavim for the immigrants, and ultimately about 200 such moshavim were founded.

Despite the numerous difficulties encountered by the immigrant moshavim, some of which were eventually abandoned, the extent of immigrant absorption in moshavim exceeded all expectations. The immigrant moshavim altered the numerical ratio between the kibbutz movement and the moshav movement. The kibbutz population, which in 1948 accounted for about half of the rural population, fell to 20 percent by 1954; the number of moshavim rose from 25 percent of the rural settlements to about 75 percent, and their population grew almost fourfold. There was also a significant change in the share of the Jewish population living on kibbutzim: from about 8 percent just before the establishment of the State down to 4 percent in 1954 (Hacohen 1994, pp. 144–145).

Was it wise to settle the immigrants on moshavim? Was their absorption in rural settlements a success story? Despite obvious differences among ethnic groups in way of life, customs, and ability to adapt, there appear to be definite examples of socioeconomic success and failure among all the ethnic groups on the immigrant moshavim (Weitz and Rokeach 1986). Most scholars believe that the failures in absorbing the immigrants in rural settlements outnumber the successes. According to Segev's study, the experience of the immigrant moshavim was by no means a success story. For many immigrants it was a tale of failure and suffering (Segev 1984). Housing conditions and the quality of public services were very poor. Many of the immigrants had no idea how to work their land and what to do with the cows and chickens that they had been given.

Indeed, studies have found that the agricultural settlements, especially the immigrant moshavim, were plagued by severe social and economic problems. Weintraub argues that there were three flaws in the planning of the rural settlements (Weintraub 1954): First, the immigrants were sent there without any emotional or ideological preparation. Second, the settlers were not screened, not even for age and family size. Third, the formation of an appropriate elite among the residents themselves was neglected. Consequently, work was neglected and crops failed. Families, especially young ones, left the moshavim. In another study, Weintraub argues,

> Because most [of the immigrant moshavim] were founded during the large immigration wave and at a time of pressure, the settlers who were sent to these moshavim underwent virtually no screening process, and they arrived without any vocational training or spiritual preparation. As a result, many of the new settlers were not able to acclimatize . . . and adjust to the role of moshav member. (Weintraub 1960)

Weingrod and Minkowicz found in their study of 27 immigrant moshavim that the residents' educational level was poor compared to that of other population groups. As a result, in the long run the settlers lacked knowledge that was essential for modern agriculture. Furthermore, they said,

> These villages are relatively closed enclaves in Israeli culture. The youth of the immigrant moshavim are not sufficiently integrated in the system of values and symbols of Israeli society and largely lack the skills needed to function more effectively in the overall framework. . . . The significance of this trend is a growing cultural gap between the rural communities and other sectors of society. (Weingrod and Minkowicz 1963, pp. 371–372)

In their opinion, this was a new phenomenon in rural Israel, which had been elitist until the 1950s.

Ra'anan Weitz (who was involved in planning the project that settled immigrants on moshavim in the 1950s) takes a different approach—and he is apparently in the minority. In his opinion,

> One should note with satisfaction that a substantial portion of the new settlers have struck roots in their moshavim. For the most part, the new immigrants have shown great willingness to adapt to economic conditions,

> and today one can point to a substantial number of settlers on a suitable agricultural and vocational level. (Weitz 1957)

Development of the Periphery

In addition to establishing urban and rural localities for the immigrants of the 1950s in the Israeli periphery, the government took other actions to develop the periphery and facilitate immigrant absorption in these regions. These include the following (Zilberger 1973):

Construction of Infrastructure
As part of an ideology that called for central control of the country's economy, the government was responsible for the economic infrastructure—roads, railroads, electrical systems, communications systems, and the water supply. By maintaining direct control and overseeing the geographical placement of these matters (and, specifically, by directing resources for infrastructure development to the periphery), the government set the stage for the geographical distribution of the population and of economic activity.

Provision of Public Services
The government provided most public services: education, welfare, and health care. By allocating large budgets for the development of public services in the peripheral regions, the government facilitated the absorption of immigrants in these areas.

Establishment of Basic Industries
The government took upon itself the tasks of developing natural resources—especially in the Southern District—and labor-intensive industries in the north and south. In this manner it created thousands of jobs for immigrants in the peripheral districts. These industries were labor-intensive, lacked diversity, and paid low wages (Gradus and Krakover 1976).

Incentives for Workers and Entrepreneurs
The government offered income-tax reductions for workers living in the periphery and substantial subsidies, loans, and grants for entrepreneurs opening businesses and factories in these regions. Among the best-known incentives for entrepreneurs is the Encouragement of Capital Investments Law, which gives entrepreneurs in peripheral regions large grants and loans with special terms.

The geographic population-dispersion policy in urban and rural localities—which was essentially a geographical dispersion of immigrants—was thus made possible by two main factors: The first, and most important, was the arrival of one million destitute immigrants who were at the mercy of the government in all areas of life. The second was the centralized economy, controlled mainly by the government, which determined not only the structure of the national economy but also the geographical distribution of the population and of economic activity.

This policy was quite successful. The proportion of the population in the Tel

Aviv and Central districts declined, and the proportion in the peripheral districts rose (for a comprehensive discussion, see Shachar 1971; Lipshitz 1996b). The ratio of the population of the Israeli core (the Tel Aviv District) to that of the periphery (the Northern and Southern districts) was 6 to 1 in 1948, and within the short period of ten years (by the 1950s) it was down to 3 to 1, where it remained for many years. The decline in the ratio was due to a high population growth rate in the peripheral districts compared to the core districts, especially the Tel Aviv District. This dramatic decline, which expresses the depolarization of the geographical distribution of the Israeli population, is a direct outcome of the Israeli population-dispersion policy. As mass immigration to Israel decreased in the early 1960s, the numerical ratio of the population of the core districts to that of the peripheral districts stabilized (for a discussion, see Lipshitz 1996b). Another example of the depolarization of the geographical distribution is that in 1948 less than one percent of the Jewish population lived in the Southern District, compared to 12 percent in 1972 (table 2 and figs. 6 and 9). The percentage of the Jewish population living in the Tel Aviv District dropped from 45 percent in 1948 to 33 percent in 1972 (compare fig. 9 with fig. 6). The standard distance (a geostatistical parameter that measures spatial dispersion), which has increased significantly since 1948, provides a good expression of the lessening of the geographical concentration of the Jewish population in Israel (Shachar 1971).

Table 2: The Geographical Distribution of the Israeli Population, by District, 1948, 1961, and 1972

	1948		1961		1972	
District	Number	Percent	Number	Percent	Number	Percent
Tel Aviv	302,100	43.2	692,600	35.9	988,900	33.5
Central	106,200	15.2	380,100	19.7	764,800	19.9
Jerusalem	84,200	12.0	187,700	9.7	346,700	9.7
Haifa	147,700	21.1	322,300	16.7	465,100	15.2
Northern	53,400	7.6	194,300	10.0	327,000	9.5
Southern	6,000	0.9	155,300	8.0	433,700	12.1
Total	716,700	100.0	1,932,300	100.0	3,350,000	100.0

2.3 The Socioeconomic Traits of the Immigrants of the 1950s and Their Spatial Impact

What were the socioeconomic traits of the immigrants of the 1950s? What effect did these traits and the government's population-dispersion policy have on the macroeconomic sociospatial (core-periphery) structure that emerged in Israel following the wave of immigration of the 1950s? We discuss these questions here.

The socioeconomic level of the immigrants of the 1950s was low, especially compared to the European immigrants who arrived in the pre-State period and to native-born Israelis. The demographic composition, level of education, and occupations of the immigrants from Asia and Africa were fundamentally different from those of the immigrants who came from displaced-persons camps in Germany, Austria, and Italy and from detention camps in Cyprus.[2]

Because the Asian and North African immigrants were from underdeveloped countries, they included a high percentage of children and a low percentage of senior citizens, and their fertility rate was high; in contrast, the Holocaust survivors who immigrated from the camps in Europe included relatively few children and a high proportion of adults (Sicron 1957). Because of the age distributions, the share of immigrants from North Africa and Asia in the labor force was lower than that of the nonimmigrant population and the immigrants from Europe.

The immigrants from Asia and Africa had a much lower educational level than the native-born population and the European immigrants. For example, 22.5 percent of the men and 57.8 percent of the women who immigrated from Asia and Africa had never been to school at all, whereas only 2.6 percent of the men and 6.3 percent of the women from Europe fit into this category. About 20 percent of the immigrants from Europe had completed post-primary education, as had 5 percent of those from Asia and Africa (Sicron 1957).

The *occupational breakdown* of the immigrants who arrived in the 1950s was fundamentally different from that of earlier immigrants and natives (Sicron 1957). Among the immigrants who had been breadwinners abroad, the largest group (41 percent of total immigrant breadwinners) consisted of artisans (e.g., tailors, shoemakers, carpenters, seamstresses, and silversmiths) and industrial workers. The second-largest

Fig. 9: The geographical distribution of the Israeli population after the immigration of the 1950s, by district (1972) [cf. fig. 6]

2. The detention camps in Cyprus held Jews who had been captured by the British Mandatory authorities while trying to enter Palestine after World War II and before Israel gained its independence.

group (about 16 percent) consisted of merchants and peddlers. Only about 5 percent of the breadwinners said they had been farmers abroad (based on Sicron 1957). The immigrants who arrived in the 1950s from Asia and Africa thus had a relatively low level of education and were concentrated in the lowest-level occupations. As a result, they entered the lowest social and occupational strata and had the lowest income levels, and gaps developed between them and the rest of the population. For example, about 15 percent of the nonimmigrants and 8.5 percent of the immigrants from Europe and America who had arrived by the 1950s worked as scientists or in liberal or technical professions, as compared with only 3.4 percent of immigrants from Asia and Africa (Sicron 1957).

Sicron notes that in terms of countries of origin, immigration after 1948 differed considerably from that in the Mandate period. "First of all, there has been an absolute and even relative increase in the number of immigrants from Asia and Africa, whereas there is relatively little immigration from Europe; and second, entire Jewish communities have been moved" (Sicron 1957; for the reasons, see his monograph). After analyzing the data on length of time in the country, Sicron concludes that "in 1948 the Jewish population of Israel was more homogeneous, relatively speaking, in that the native-born accounted for over 35 percent, and about 65 percent of the foreign-born had lived [in Palestine for] at least ten years" (Sicron 1957). It seems that this demographic change is the key to understanding the geographical, economic, and social changes.

In the pre-State waves of immigration, the immigrants themselves were an important source of two factors: they brought most of the capital (at least since the early 1930s), and they had acquired a fairly high level of education in their countries of origin. Most of the post-1948 immigrants were destitute and much less educated and skilled than the general Israeli population. The differences in level of education, technological skills, and occupations between recent and not-so-recent immigrants from Europe and the new arrivals from North Africa and Asia was evident in income levels, housing conditions, and consumption patterns. In his study of the Israeli economy during the first decade of the State of Israel, Patinkin explains,

> The average income of the old-timers is higher than that of the new immigrants and [the average income] of those from Europe-America [is higher] than that of [the immigrants] from Asia-Africa. Because families born in Asia-Africa are larger, on average, than the European families, the difference in per-capita income is even greater. . . . Thus there is no sign whatsoever of a reduction in the income disparity between the old-timers and the new immigrants after the immigrants have been in Israel for a while.

As a matter of fact, there are indications that this gap became even wider (Patinkin 1959, pp. 61–63).

In terms of consumption patterns, too, a substantial disparity emerged between the immigrants who arrived in the 1950s and the veteran Israeli population. In his study of consumption patterns in Israel, Levitan (1964) argues that there were distinct differences between Israelis from the different continents in consumption of education, culture, and health services; European emigrants used these services 14, 90, and 70 percent more, respectively, than did Asian emigrants. Furthermore, the disparities between continents did not decrease overall the longer the immigrants had been in Israel. Educational disparities actually *increased*, because, although the Asian population tended to restrict its expenditures on clothing and footwear and increase spending on education, culture, and health care, European emigrants also spent more on education and culture, and therefore the "net" result was sometimes an enlargement of the gap in consumption patterns (Levitan 1964).

The mass immigration of the 1950s produced the new state's first substantial income disparities. In his comprehensive study of income inequality in the 1950s, Hanoch states,

> Until the establishment of the State, Israel had one of highest degrees of equality of all countries in the world. . . . Since the establishment of the State, things have changed considerably. The organized pioneering immigration gave way to a more diverse mass immigration, both in terms of level of education and occupational training and in terms of social and economic background. Pioneering values and the egalitarian ideology lost importance. The relative abundance of people with medium and high levels of education and expert professionals was replaced by a fairly severe shortage, and the need to encourage large numbers of people to enter high-level occupations induced wider gaps in the financial remuneration paid to the groups on high educational and occupational levels. . . . Consequently, a substantial social and economic gap developed between the veteran population of the country—whose economic position was established over the years and who formed the class that held the major posts in the economic and social institutions—and the new immigrants, who were mostly destitute and in need of support, and many of whom were uneducated and lacked technical training on a level suitable for immediate, full economic and social absorption in the modern, developing Israeli economy. (Hanoch 1961, pp. 35–36)

In this context Hanoch shows that in the 1950s there were substantial differences in the income of breadwinners from different countries, and that the disparity in standard of living (per-capita income) between those from the West (Europe and North America) and those from the East (Asia and North Africa) was even greater than the income disparity among breadwinners, because the members of Oriental ethnic groups had larger families and thus a greater number of children and other dependents to support.

The educational differences between the earlier immigrants—natives of Europe and America—and the immigrants of the 1950s—natives of Asia and Africa—have wide-ranging effects, including in the distribution of employment. Bar-Yosef and Padan (1976) show that in the 1950s there was a high concentration of European and American immigrants in prestigious occupations, whereas the immigrants from Asia and Africa were virtually unrepresented. They explain, inter alia, that the immigrants from non-industrialized Arab countries remained on the low rungs of the class ladder in Israel because, at least in the 1950s and 1960s, they did not adapt to the class structure of an industrial society (Bar-Yosef and Padan 1976).

Government activity in the peripheral regions with respect to housing, employment, and public services—as described at the beginning of this chapter—did not attract many long-time residents from the core districts to the periphery, but they did manage to cause immigrants to settle initially in the periphery. This—and not internal migration by the nonimmigrant population—is what transformed the geographical distribution of the Israeli population, and of the Jewish population in particular: from a polarized spatial structure to a graduated spatial structure (for details, see Shachar 1971 and figs. 6 and 9).

Between 1948 and 1972 the proportion of the Jewish population in the core districts declined, and that in the peripheral districts rose. For example, the proportion of the Jewish population in the Tel Aviv District fell from 43 percent in 1948 to 33 percent in 1972 and the proportion in the Southern District rose from 0.9 percent in 1948 to 12 percent in 1972.

Data from the Central Bureau of Statistics also show that a higher percentage of European immigrants than Asian and African immigrants settled in the Tel Aviv and Haifa districts (about 57 percent of European immigrants as opposed to 35 percent of Asian and African immigrants). A larger share of Asian and African immigrants settled in the Northern, Southern, Jerusalem, and Central districts. As a result, at the end of 1953 immigrants from Asia and Africa accounted for 43 and 62 percent of the populations of the Northern and Southern districts, respectively, but only 27 percent of the population of the Tel Aviv District and 24 percent of that of the Haifa District (Sicron 1957).

As described earlier in this chapter, the immigrants who arrived in Israel in the 1950s were on a low socioeconomic level. Consequently, the dispersion of the immigrants to the peripheral districts turned the geographical periphery into a socioeconomic periphery, too, although it had not been one previously. *Thus, the population-dispersion policy for the immigrants of the 1950s lessened the geodemographic polarization (in terms of the number of residents) but simultaneously—mainly because of the immigrants' traits—created a spatial-socioeconomic polarization (in terms of the quality of human capital) between the core and the periphery* (Shachar and Lipshitz 1981; Lipshitz 1986b, 1991). The core districts have a fairly high proportion of European- and American-born inhabitants, and they score higher in terms of "well-being and development variables" (e.g., higher education, income level, motorization rate, and housing conditions); whereas the peripheral districts have a relatively high percentage of Asian- and African-born inhabitants, who score low in terms of these variables (for an example, see fig. 10).

Figure 10 indicates that in the 1950s there was a positive spatial correlation between the socioeconomic well-being of a region and the percentage of natives and pre-State immigrants in it; or alternatively, there was a negative spatial correlation between the socioeconomic well-being of a region and the percentage of recent immigrants in it. This idea is supported statistically by regressing several independent variables, including length of time in the country, on economic well-being. The analysis shows that in the early 1960s the geographical variation in length of time in Israel was the most important factor underlying the geographical variation in economic well-being (Lipshitz 1982, 1984). A multivariate regression analysis of the 1961 census data indicates that length of time in Israel contributed about 51 percent of the explanation of the

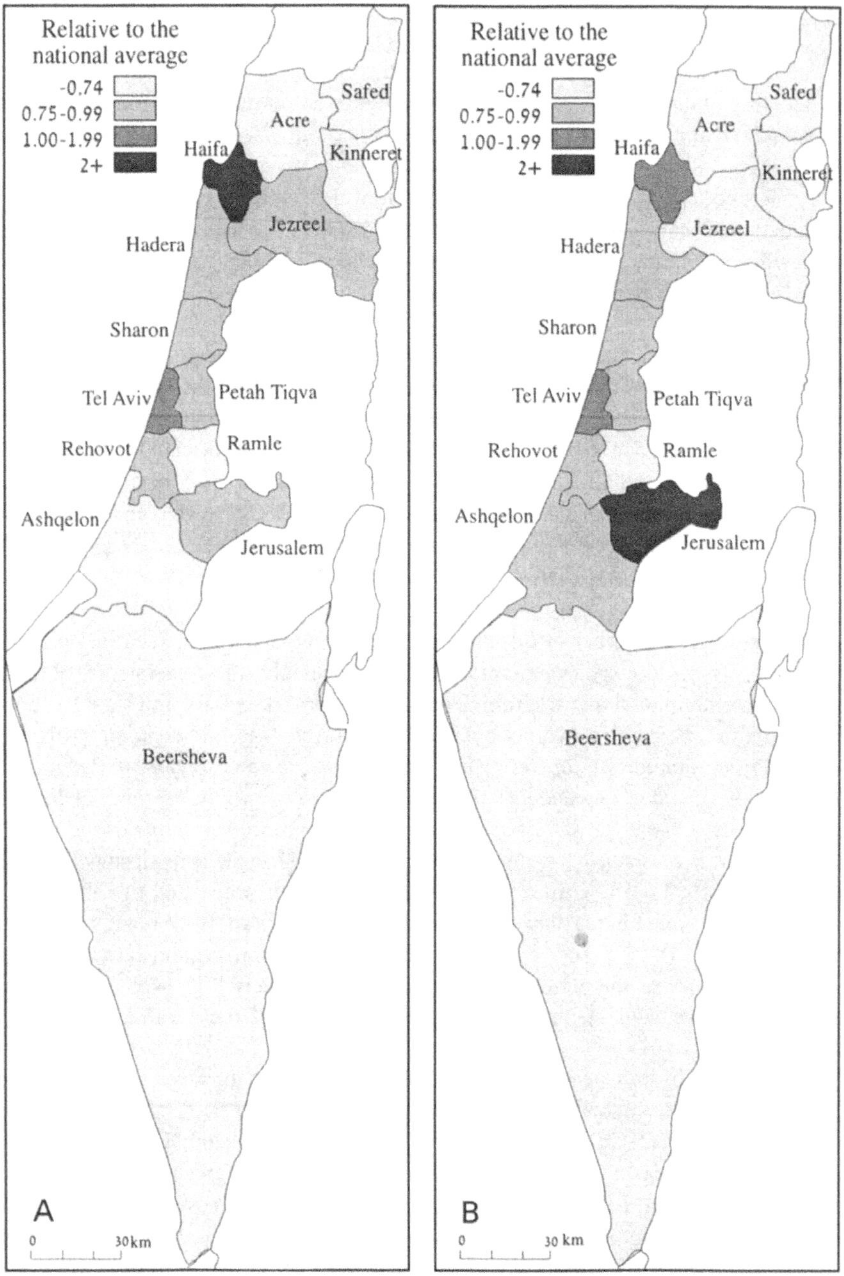

Fig. 10: The geographical distribution of the Israeli population, 1961: (A) by higher education; (B) by the ratio of European and American immigrants to Asian and African immigrants (by subdistrict)

socioeconomic differences between locations in Israel, and the other variables contributed about 30 percent. In 1961 the spatial correlation between length of time in Israel and the percentage of immigrants from Asia and Africa was –0.84. In other words, in the 1960s and 1970s the length of time a person had lived in Israel—which was an indication of the person's country of origin—influenced his or her socioeconomic level, and the distribution of the immigrants by length of time in Israel (i.e., by country of origin) determined the geographical distribution of affluence and poverty.

To sum up, the map of economic well-being that took shape in Israel in the 1950s is a spatial result of the encounter between two categories of immigrants—pre-State and post-1948—and proved to be stable in the long term (Shachar and Lipshitz 1981; Lipshitz 1986b). In other words, regions that were ranked high (or medium or low) on the Israeli well-being scale in the 1950s were still in the same relative location in the 1980s. The map of economic well-being can be altered only by a major shock of change. The shock that for many years determined the spatial distribution of the affluent population in Israel was the settlement process of the pre-State immigrants; their affluence is partly due to an "import of development" and partly the result of their economic success in the pre-State period. The shock that determined the spatial distribution of the disadvantaged population in Israel was the wave of immigration of the 1950s; their poverty is largely due to an "import of underdevelopment."

The spatial distribution of nonimmigrants and of European and American immigrants who arrived in the 1930s and 1940s, combined with the spatial distribution of the immigrants who arrived in the 1950s, created the spatial-socioeconomic (core-periphery) structure in Israel.

At the end of the 1950s immigration to Israel declined dramatically. The social and economic disparities between core and periphery that took shape in the late 1950s are the main factor in the other kind of population movement in Israel: internal migration between Israeli regions. This will be discussed in the next chapter.

CHAPTER 3

INTERNAL MIGRATION AND REGIONAL DISPARITIES IN THE CORE-PERIPHERY STRUCTURE

3.1 Introduction

The previous chapter showed that the geographical distribution of immigrants in the pre-State period—most of them from Europe—determined the distribution of the affluent population in Israel; and that the geographical distribution of the immigrants of the 1950s determined the distribution of the disadvantaged population. The high level of well-being of the former is partly due to the "import of development" by the immigrants and partly the result of their economic success in Palestine in the 1930s and 1940s. The economic hardship of the immigrants of the 1950s is mainly the result of their "import of underdevelopment," which is related to their economic, educational, and occupational traits in their countries of origin.

The previous chapter also showed that the geographical distribution of the immigrants of the 1950s significantly reduced the disparity between the core and the periphery in terms of the number of inhabitants. But what about the spatial distribution of level of well-being/development in Israel? More specifically, what spatial pattern describes the socioeconomic disparities that developed between the core and the periphery as a result of the geographical dispersion of the immigrants of the 1950s, and how does this pattern vary over time? What spatial pattern does internal migration follow in Israel, and how does it vary over time (in the 1960s, 1970s, and 1980s)? How is the model of internal migration affected by socioeconomic disparities between regions, and how does it affect these disparities?

In terms of the time dimension, these questions refer to the 1960s (especially the second half of the decade), 1970s, and 1980s. *During these decades internal migration replaced immigration as the major determinant of the geographical distribution of the population in quantitative and qualitative terms* (the latter being the socioeconomic traits of the population, which are a major determinant of well-being). Internal migration is a manifestation of the geographical flow of market forces more than of government policy. We know from the international professional literature that internal migration generally flows from the periphery to the core (Vining and Pallone 1982; Brown and Lawson 1989). The geographical distribution of the population—which is determined by internal migration—has a direct impact on the distribution of housing, businesses, jobs, and infrastructure (and vice-versa). The population distribution reflects the spatial macroeconomic structure (e.g., the core-periphery structure) more than government spatial policy. In a certain sense it can be argued that internal migration and government activity generally work in opposite geographical directions: the former from periphery to core, and the latter from core to periphery.

The primary aim of this chapter is to examine the geographical patterns of internal migration in Israel in the period between the two peaks of immigration in the 1950s (when immigrant absorption was controlled by government spatial policy) and the 1990s (when the government spatial policy is to let immigrant absorption be based chiefly on market forces). The spatial patterns of internal migration are affected by a number of factors, but from theoretical and empirical studies (Sonis 1980; Brown and Lawson 1989; Vining and Pallone 1982; Lipshitz 1991a, 1991b) we know that regional socioeconomic gaps (disparities between regions in level of well-being) are the main factor. Therefore we will look first at the spatial patterns of these disparities, which took shape—as described above—in the geographical encounter between the immigrants of the 1950s and those of the 1930s and 1940s, and then we will consider internal migration as a reaction to the regional disparities.

3.2 Methodological Aspects: The Research Variables and the Boundaries of the Research Areas

Two basic methodological problems in the present section are what variables to use to measure regional economic well-being and how to determine the boundaries of the research areas. These are the two fundamental problems in any research study that examines changes in regional disparities in well-being. Selection of the variables for measuring well-being and demarcation of regional boundaries have direct implications for the results of the study; moreover, the boundaries must be kept constant so that comparative studies over time are possible. This section will not discuss these problems (on the issue of research areas, see Lipshitz 1986a). We merely note that in empirical studies we find two main approaches to measuring well-being in general, and regional well-being in particular: the *objective approach*, which relies on quantitative variables such as per-capita income, level of education, housing conditions, level of unemployment, and consumption variables; and the *subjective approach*, which measures well-being based on the residents' perceptions and opinions (for a comparison of the approaches, see Lipshitz 1993b). Whereas the objective approach relies mainly on statistical publications and does not require direct contact with the research population, the subjective approach is more anthropological ("participant observation") and requires direct contact between the researcher and the research population (questionnaires, interviews, surveys, etc.) to evaluate subjective well-being.

It is commonly believed that there is a positive correlation between financial status and satisfaction with quality of life, but several studies show that this link is not so clear. In an empirical study, Kuz demonstrates that the correlation between quantifiable objective variables such as standard of living and subjective variables such as a sense of quality of life is low, and that to characterize the regional standard of living one must use both kinds of variables (Kuz 1978). Most studies use only the objective type and suggest numerous variables that can be used to measure the regional standard of living.

The variable preferred by most researchers who favor the objective approach is regional income per capita (for a comprehensive discussion, see Richardson 1978). They argue that all the other "objective variables" affect or are affected by real income per capita. In statistical terms, they argue that there is a high correlation—whether positive or negative—between per-capita income and each of the objective variables. For example, consumption variables and housing conditions are influenced by real family income (or by the income of each family member), whereas one's level of education and occupation affect one's income level. Thus, when one assesses level of well-being based on real per-capita income, one has already taken all the other relevant variables into account. It should be noted that some researchers disagree with this approach, saying that well-being is multifaceted and that, therefore, a variety of measurable "objective variables" should be selected, such as income, education, crime rate, consumption, air pollution, unemployment, and public services (for a discussion of this subject, see Lipshitz 1993b).

Because data on the regional distribution of per-capita income over time are not available in most countries, as well as for the reasons mentioned in the previous paragraph, researchers use other variables for which data by geographical unit are available; some examples are housing conditions (Hillard et al. 1972), ownership of durable goods (Coates, Johnson, and Knox 1977), percentage employed in agriculture (Williamson 1965), variables related to quality of life and the environment (Smith 1979; 1994), motorization rate (Slater 1975; Shachar and Lipshitz 1981), and level of education (Smith 1979).

In Israel, data on the distribution of per-capita income by region are available only for 1972 and 1983, when population and housing censuses were conducted. Data on the per-capita income of wage-earners by locality (not region) have been published by the National Insurance Institute since the early 1990s. Because these data come from a different source and were obtained in a different manner, they cannot be compared to earlier data. Studies in Israel that have examined possible alternatives to per-capita income have found that the rate of private-vehicle ownership per 1,000 population (hereinafter: motorization rate) and the percentage of the population with higher education are good alternatives to per-capita income for the purpose of measuring regional economic well-being (Reichmann, Eliash, and Berashi 1971; Shachar and Lipshitz 1981). In keeping with the availability of data in Israel, regional economic well-being is measured in the present study in terms of the motorization rate for 1962, 1972, and 1982 and in terms of higher education for 1961, 1972, and 1983.

The advantage of using these variables is that the figures for each Israeli region have been available since the early 1960s. The disadvantage is that their validity over time—especially that of the motorization rate—as a substitute for regional income per capita is decreasing. Because the present chapter focuses on the 1960s, 1970s, and mid-1980s (the period between the two peaks of immigration to Israel in the 1950s and the 1990s), this drawback is less significant. Nevertheless, where appropriate and when the data permit, we will include some figures on per-capita income, too.

With respect to the research areas (the second methodological problem), this study, like many others, is forced to be more realistic than idealistic and to use standard administrative regions, the only spatial units for which time-series data are available. It should be noted that in spatial-economic research it is best to use spatial-economic units that reflect the boundaries of the labor markets of cities of various sizes. These boundaries vary over time, making it difficult to carry out a comparative spatial-economic study over time. Moreover, as was mentioned above, in many countries the published statistical data are based on administrative geographical units, e.g., provinces in Canada, states in Brazil, and districts and subdistricts in Israel. Consequently, many studies—including this one—are forced to consider trends of regional disparities and internal migration on the basis of administrative units whose boundaries have remained constant. This, of course, has its advantages and disadvantages, as was described above briefly (for a discussion of this in connection with migration, see Champion 1988; Lipshitz 1996b).

The research areas used are therefore the natural regions, districts, and subdistricts of the State of Israel. Israel is divided administratively into 6 districts, 14 subdistricts, and 40 natural regions (see figs. 11, 12, and 13). The study also refers to six macro regions that make up the spatial-economic core-periphery structure of the country (based on Shachar and Lipshitz 1981; Lipshitz 1996b). Each of these six regions comprises several natural regions, as follows (see fig. 14):

1. Primary core	Tel Aviv region (natural region no. 30)
2. Secondary core	Haifa region (natural region no. 19) Jerusalem region (natural region no. 1)
3. Tel Aviv periphery	(natural regions 25, 26, 27, 28, 29)[1]
4. Intercore region	Northern (natural regions 21, 23, 24) Southern (natural regions 33, 34) Eastern (natural regions 2, 27)
5. Tertiary core	Beersheva region (natural region no. 37)
5. National periphery	Northern (natural regions 3, 4, 5, 6, 7, 8, 11, 12, 13, 14, 15, 16, 17, 18) Southern (natural regions 31, 35, 36, 39, 40)

The "primary core," the Tel Aviv district, the Tel Aviv subdistrict, and the natural region of Tel Aviv are coterminous. The "secondary core" consists of the Jerusalem and Haifa natural regions. The "Tel Aviv periphery" contains the natural regions of Rehovot, Rishon Lezion, Petah Tiqva, and the southern Sharon. The "intercore region" consists of most of the coastal plain—Nahariyya, Acre, Netanya, Hadera, Zikhron Ya'aqov, Ashdod, and Ashqelon—as well as the Lod region and the Judean foothills. The "tertiary core" consists of the Beersheva region. The "national periphery" includes most of the natural regions in the Northern District (the Galilee) and the Southern District (the Negev).

This division applies to Israel of the 1960s and 1970s, and perhaps even the early 1980s. It is an "average" geographical division (if the word "average" can even be used in this context) of Israel into core and periphery based on socioeconomic traits (see Shachar and Lipshitz 1981; Lipshitz 1986a). By the late 1980s, and even more so by the early 1990s, the core-periphery division of the country looked different; we will discuss this further in the conclusion. Moreover, even during the period of almost three decades

1. The Tel Aviv periphery is also referred to in this book as the "periphery of the primary core." This region includes urban and rural suburbs of Tel Aviv.

1. Jerusalem Region
2. Judean Foothills
3. Hula Valley
4. Eastern Upper Galilee
5. Hatzor Region
6. Kinnerot
7. Eastern Lower Galilee
8. Bet She'an Valley
9. Harod Valley
10. Ramat Kokhav
11. Jezreel Valley
12. Ramat Menashe
13. Nazareth-Tir'an Mountains
14. Western Lower Galilee
15. Yehi'am Region
16. Eilon Region
17. Nahariyya Region
18. Acre Region
19. Haifa Region
20. Carmel Coast
21. Zikhron Ya'aqov Region
22. Mount Alexander
23. Hadera Region
24. Sharon
25. Southern Sharon
26. Petah Tiqva Region
27. Lod Region
28. Rehovot Region
29. Rishon Lezion Region
30. Tel Aviv Region
31. Malakhi Region
32. Lakhish Region
33. Ashdod Region
34. Ashqelon Region
35. Gerar Region
36. Besor Region
37. Beersheva Region
38. Northern Arava
39. Southern Arava
40. Negev Highlands

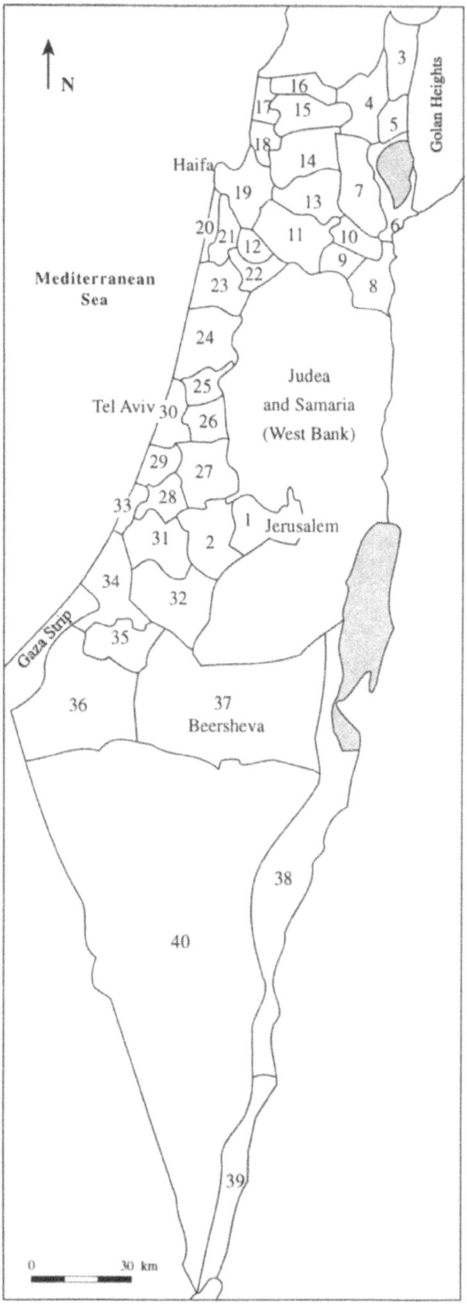

Fig. 11: The natural regions of Israel

Fig. 12: The districts of Israel

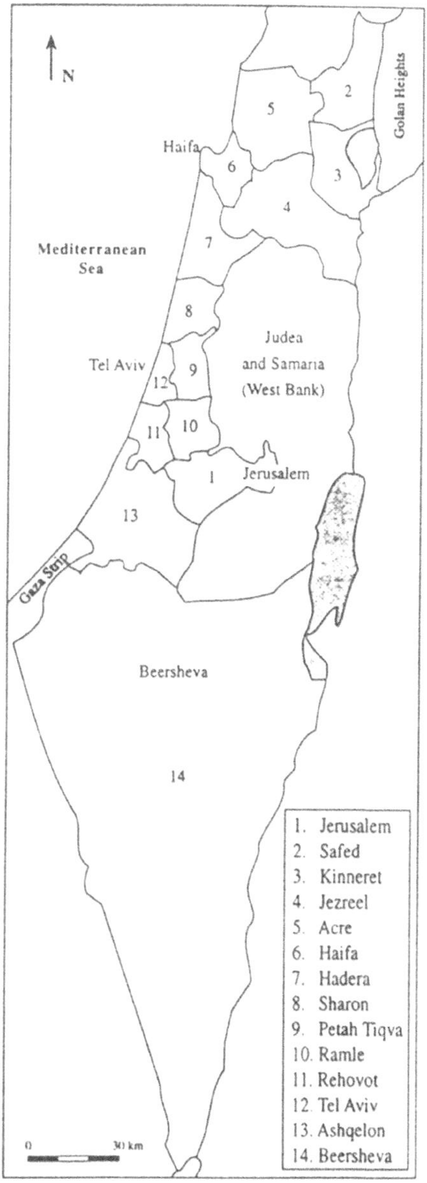

Fig. 13: The subdistricts of Israel

during which the division remained fairly constant—making it possible to conduct a comparative study over time—certain changes took place. For example, although the Beersheva region was a tertiary core in the 1980s, it was less so in the 1960s and 1970s, when Beersheva was a central city of the Israeli national periphery. The Jerusalem region became a secondary core only after 1967 (the Six-Day War), when the city was reunited and vast financial resources were put into its physical and economic development. Structural changes in the economy of the Tel Aviv metropolitan area that cause rapid growth of the core region and its periphery (Kellerman and Krakover 1986) are altering the boundaries of the primary core and its periphery.

Fig. 14: The geographical division of Israel into core and periphery. This division is obtained by combining several natural regions and several districts and subdistricts. It holds true until the mid-1980s. Since then, however, the core has spread out in the direction of the periphery (see Conclusion).

3.3 Socioeconomic Well-Being in the Spatial Core-Periphery Structure (Regional Disparities in Israel)

The analysis in this section of the spatial distribution of the level of well-being in Israel in the 1960s, 1970s, and early 1980s focuses—as was explained in the previous section—on two main variables: higher education and motorization rate. Occasionally it looks at per-capita income for 1983 and the geographical division of Israel into core and periphery, as was described in section 3.2. The present section will not analyze the spatial distribution of socioeconomic well-being (regional disparities) in depth and in detail for the following reason: The purpose of the book is to describe and analyze immigration to Israel and internal migration within Israel. This section presents only the spatial-socioeconomic results of the wave of immigration to Israel in the 1950s and serves as a basis for understanding the internal migration that ensued. The discussion is based on figures 15 and 16, which depict the level of relative well-being in each region (V) in the years 1961 (or 1962), 1972, and 1983.[2]

The spatial pattern of economic well-being in Israel in the early 1960s was a polarized one (see fig. 15 [pp. 68–69]). The motorization rates show a marked difference between the Tel Aviv and Haifa core areas, where economic well-being exceeded the national average (V>1), and the other areas, where it was below the national average (V<1). In the early 1960s the latter group was made up of two main subgroups: One comprised the Jerusalem region, the Tel Aviv periphery, and the northern coastal plain (between the Tel Aviv core and the Haifa core), where relative economic well-being was between 0.5 and 1.0. The other contained the Northern District, the Southern District, and the southern coastal plain (Ashdod and Ashqelon), which were characterized by relatively low levels of economic well-being (V<0.5) and which constituted the national periphery of Israel.

For higher education, too, we find a polarized spatial pattern in the early 1960s that resembles the pattern of the motorization rate. The main difference is that the Jerusalem region (instead of the Tel Aviv region) scores highest on the regional scale of economic well-being (see fig. 16 [pp. 70–71]). Jerusalem "produces" a population with a high rate of higher education and attracts a highly educated population for two main reasons: the presence of the Hebrew University—which until the 1960s was the only university in Israel and is still considered the most prestigious of the six Israeli universities—and the presence of a large public sector, which includes government ministries, the Jewish Agency, courts, and more.

Between 1962 and 1972, the motorization rate in Israel rose from 18.86 per thousand population to 71.33 per thousand population, and between 1961 and 1972 the share of the population with higher education increased from 9.94 to 12.55 percent. This is clear evidence of a rise in national economic well-being.

These changes on the national level were reflected by several significant changes on the regional level: the Beersheva region moved from the group with a low level of relative well-being (V<0.5) to the group with a medium level of relative well-being (0.51<V<1.00); and the relative well-being of the Tel Aviv region declined from above 1.50 to between 1.00 and 1.50.

In terms of higher education, the main changes between 1961 and 1972 took place in the areas south of Tel Aviv: the Rehovot region (part of the Tel Aviv periphery)

2. If well-being is measured in terms of higher education, then

$$V = \frac{\text{level of well-being in the region}}{\text{level of well-being in Israel}} = \frac{\text{rate of higher education in the region}}{\text{rate of higher education in Israel}}$$

If well-being is measured in terms of motorization rate, then

$$V = \frac{\text{level of well-being in the region}}{\text{level of well-being in Israel}} = \frac{\text{motorization rate in region}}{\text{motorization rate in Israel}}$$

joined the group with the highest level of relative well-being, and the Ashdod and Ashqelon regions (part of the national periphery) moved from the category of low relative well-being ($V<0.5$) to medium relative well-being ($0.5<V<1$). Between 1972 and

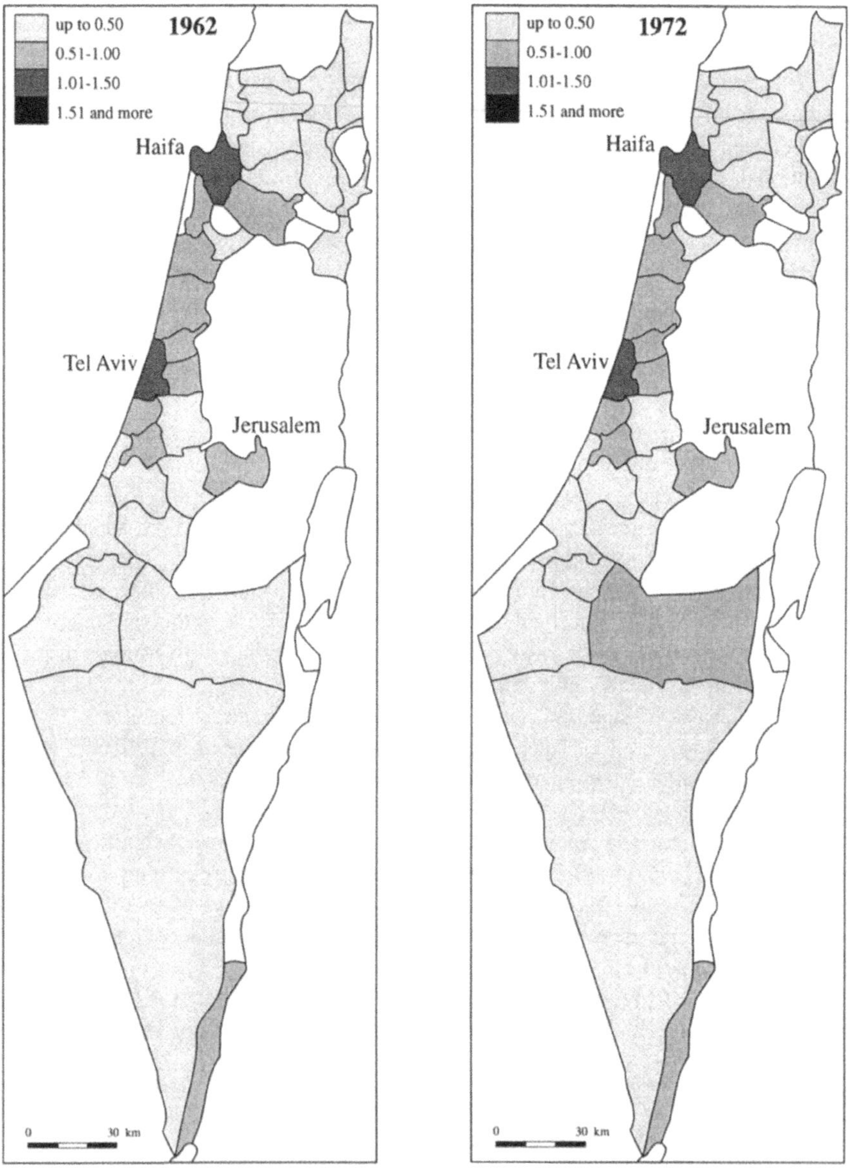

Fig. 15a: The spatial distribution of relative well-being based on motorization rate, 1962 and 1972
(national average = 1)

1983 the most notable changes took place in the Tel Aviv periphery and in the Jerusalem region. The relative well-being of the Jerusalem region increased and it joined the group of regions with high relative well-being; the relative well-being of the Petah

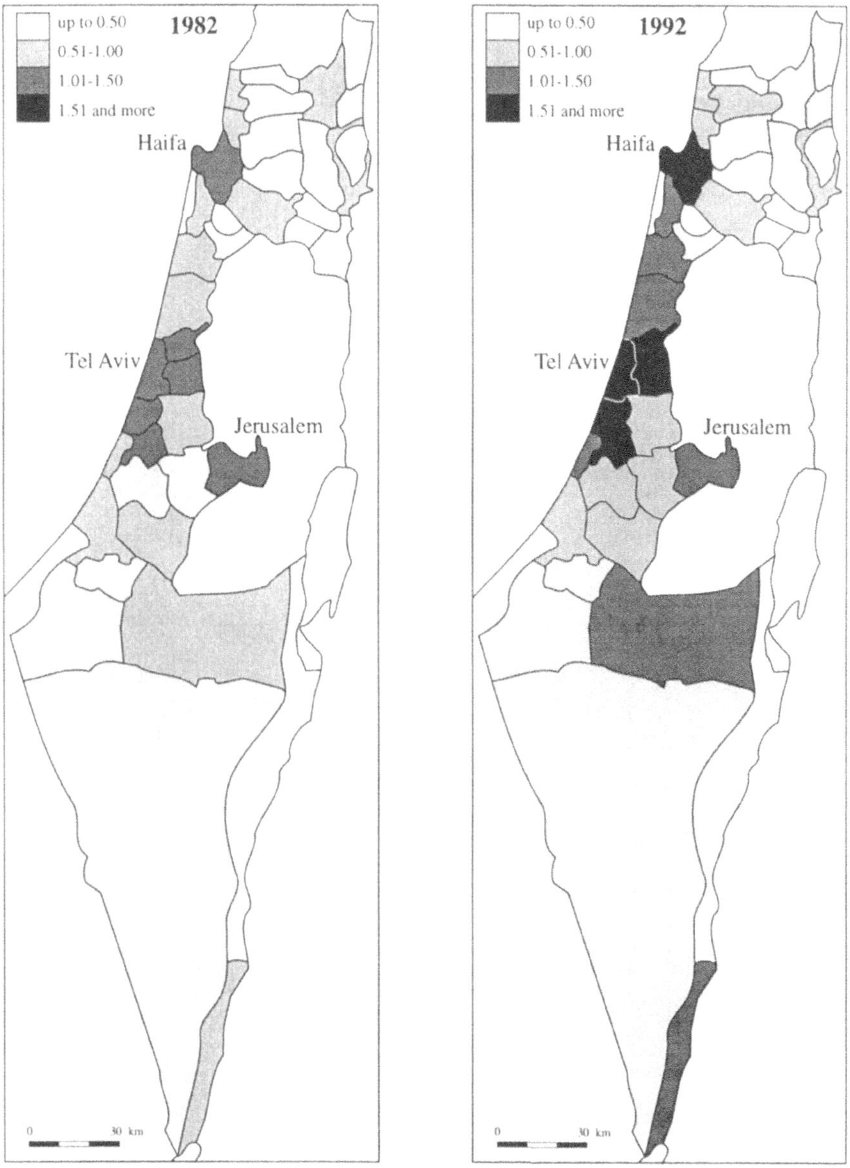

Fig. 15b: The spatial distribution of relative well-being based on motorization rate, 1982 and 1992
(national average = 1)

Tiqva, southern Sharon, and Rishon Lezion regions (all of them in the Tel Aviv periphery) also rose, and they moved from the category of medium relative well-being to that of high relative well-being.

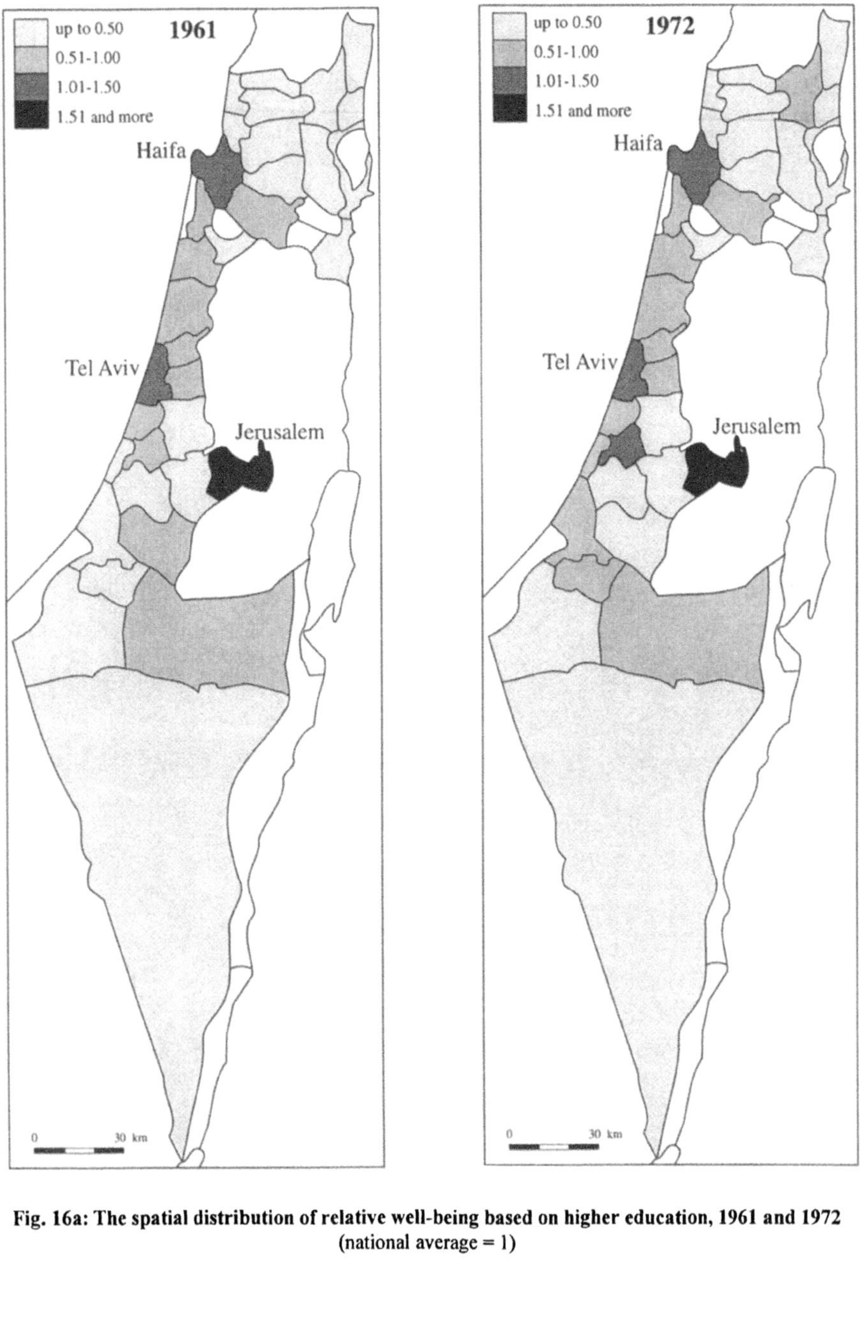

Fig. 16a: The spatial distribution of relative well-being based on higher education, 1961 and 1972
(national average = 1)

More substantial spatial/regional changes occurred during the national economic growth of the 1970s. These changes took place around the Tel Aviv core and Tel Aviv periphery, in the Jerusalem region, along the southern coastal plain (the Ashdod and Ashqelon regions), and in the Northern and Southern districts. The main changes (based on motorization rate) are as follows:

- The relative economic well-being of the Jerusalem region increased, and by the 1980s it was in the group with the highest relative well-being.
- The relative well-being of the regions in the Tel Aviv periphery increased, and by the 1980s they were in the group with the highest relative economic well-being. This change equalized the level of relative well-being in terms of the concentration indices in the periphery of the Tel Aviv region and in the Tel Aviv region itself. Thus in the early 1980s the Tel Aviv area (which included the Tel Aviv and Central districts) was uniform in terms of economic well-being.
- The relative economic well-being of the Safed, Tiberias, and Lakhish regions rose; these regions are part of the national periphery.
- The relative economic well-being of the Ashqelon and Ashdod regions rose. Both of these are in the intercore area on the southern coastal plain. This was an early sign that the Tel Aviv core was spreading southward toward the northern parts of the southern periphery (Southern District). Indeed, by the 1990s planners were treating Ashqelon and Ashdod as part of the remote ring of the Tel Aviv core.

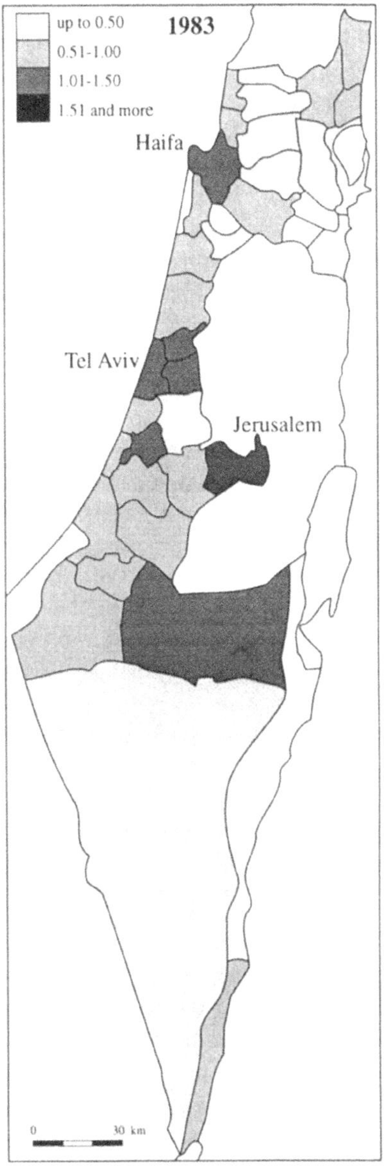

Fig. 16b: The spatial distribution of relative well-being based on higher education, 1983
(national average = 1)

The main changes that occurred between 1982 and 1992 (fig. 15) are as follows: an increase in the relative economic well-being of the Tel Aviv and Haifa cores (to V>1.51); extension of the highest category of relative economic well-being (V>1.51) to the periphery of the Tel Aviv core; creation of a high level of relative economic well-being ($1.01<V<1.50$) in the distant suburbs of the Tel Aviv metropolitan area (the Sharon region in the north and Ashdod in the south); extension of the high level of relative economic well-being toward regions in the periphery of the Haifa metropolitan area, such as the western Lower Galilee; and an increase in the relative economic well-being of regions located on the "seam" between the national periphery and the periphery of the Tel Aviv metropolitan area. Even if the motorization rate is of limited validity in describing the economic well-being of regions in the 1990s, these findings are evidence of the extension of the Tel Aviv and Haifa core regions toward areas that were once part of the intercore regions and the national periphery. This extension is a manifestation of deconcentration processes on the metropolitan level.

Above we described the main changes in socioeconomic well-being in the various regions in Israel. Some of the changes are a quantitative expression of the move from one class interval to another. Are these changes indeed significant? Are the geographical patterns described above stable over time? We examine this question by means of a regression analysis between the regional distributions of the two research variables: motorization rate for 1962, 1972, and 1982 and higher education for 1961 and 1972 (Lipshitz 1986b).

Analysis of pattern stability is carried out by analyzing the regression parameters for the welfare distribution by region in period t and the distribution in period $t-n$. The regression analysis is based on an examination of the values of two parameters: the correlation coefficient between the above distributions and the slope of the regression line. The closer the correlation coefficient for the regression between the t and $t-n$ distributions is to 1, the higher the level of stability is in the spatial patterns between the periods. In other words, the hierarchical structure of regional economic welfare was being preserved. On the other hand, if the correlation were closer to –1 this would signify that between the periods an almost complete upset had taken place in the spatial pattern of economic welfare; however, in human geography such instability would be quite extraordinary.

Even if a high level of stability is established, it is important to know the extent to which the various regions contribute to this general stability. Examination of the values of the slope of the regression line may help supply a partial answer. If, in the examination of the connection between the t and $t-n$ welfare distributions, a regression slope greater than 1 is obtained, then it would be held that, despite the stability, the relatively strong regions gained even more strength in comparison with the weaker regions between the two periods. However, if the slope was less than 1, it could be claimed that the stronger regions grew weaker in relative terms, while the weak ones gained strength.

In the light of this introduction, it is now possible to examine the stability of the spatial distribution of socioeconomic well-being in Israel. Using motorization as the welfare criterion, it is possible to assess stability between the early 1960s (1962) and 1970s (1972) and between the 1970s and early 1980s (1982). For the education criterion, however, only the earlier period can be analyzed. The results are shown in table 3.

The findings of the regression analysis show that the correlation for the years studied is positive, high, and significant. In other words, the regions with low levels of

Table 3: Parameters of the Regression Analysis between the Variables Studied

	Motorization rate variables		Higher education	
Parameters	1962–1972	1972–1982	1962–1982	1962–1972
Correlation coefficient (*r*)	0.93	0.95	0.88	0.87
Intercept (*a*)	10.40	37.70	55.23	2.54
Slope (*b*)	3.59	1.67	5.98	0.94

economic well-being in 1962 were marked by low levels of economic well-being in 1972 and 1982 as well. The same is true of the regions with other levels of economic well-being.

The significance of these findings is that the regional disparities that developed in Israel during the 1950s as a result of the geographical distribution of the pre-State waves of immigration and the wave following independence (in the 1950s)—as was described in detail in chapter 2—have remained stable over time. As stated, in the early 1960s immigration to Israel declined greatly, and the main population movement shaping the socioeconomic and even physical spatial pattern was *internal migration.* This movement was highly influenced by regional disparities, as will be seen in the following sections.

3.4 Internal Migration within the Core-Periphery Structure (1961–1988)

The purpose of the present section is to examine internal migration in Israel within the spatial economic framework of a core-periphery structure during the years 1961–1988. The geographic division of Israel into a core-periphery structure actually reflects regional differences in level of well-being, as was described in the previous section (for a broader discussion of this division, see Gradus 1983; Lipshitz 1986a). The period from 1961 to 1988 was between two peaks of immigration: the 1950s and the 1990s. During this time internal migration was the main population movement affecting the economic, social, and physical spatial patterns in Israel.

NET MIGRATION WITHIN THE CORE-PERIPHERY STRUCTURE IN ISRAEL

Internal migration in Israel—as elsewhere in the world—is greatly affected by the spatial distribution of economic well-being described in the previous section; this subject will be addressed in the rest of the chapter. The discussion of internal-migration trends in Israel begins with a description and analysis of changes over time in net migration in the spatial framework of core-periphery. Net migration (the number of in-migrants minus the number of out-migrants in each region) essentially reflects how attractive a region is, and one of the main factors in attractiveness is the region's level of well-being (level of development).

Examination of the trends of the spatial pattern of net migration between 1961 and 1988 reveals that the pattern in the Tel Aviv region (the country's primary core) underwent significant changes during this period (see figs. 17a and 17b). From being the major beneficiary of population movements with the highest positive migration balance until the late 1960s, the Tel Aviv region shifted to a negative migration balance in the mid-1970s. This shift in net migration shows the dramatic changes that took place in the Tel Aviv core area during the 1960s and 1970s, when the virtual saturation of the core in terms of built-up area, the high costs of housing, competition for apartments in the city between businesses and prospective residents, crowded roads, and serious ecological problems (pollution and noise) increasingly undermined and reduced the power of attraction of the Tel Aviv core. Concurrently, the power of attraction of the Tel Aviv periphery (urban and rural suburbs of Tel Aviv), where housing was less expensive than in the Tel Aviv core and ecological conditions were better, rose. Whereas in 1961 this latter region had a negative net migration, beginning in the mid-1960s the balance swung to the positive, and the values of this positive balance increased steadily with time.

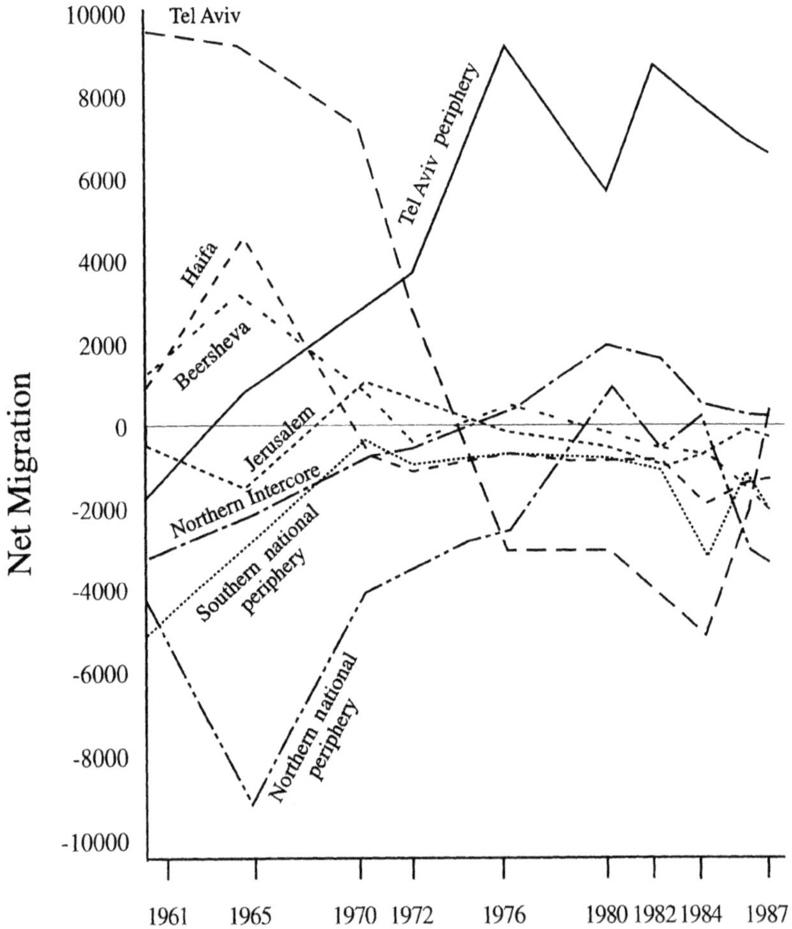

Fig. 17a: Net migration within the core-periphery structure, 1961–1987 (absolute numbers)

We may note that from the mid-1980s on, there is evidence of a slowdown in the negative net migration in the Tel Aviv region. In 1987, the negative balance swung over to the positive side, reflecting the onset of a process in the Tel Aviv region (as in other metropolitan areas of the world) termed "back to the city." Migration to the Tel Aviv

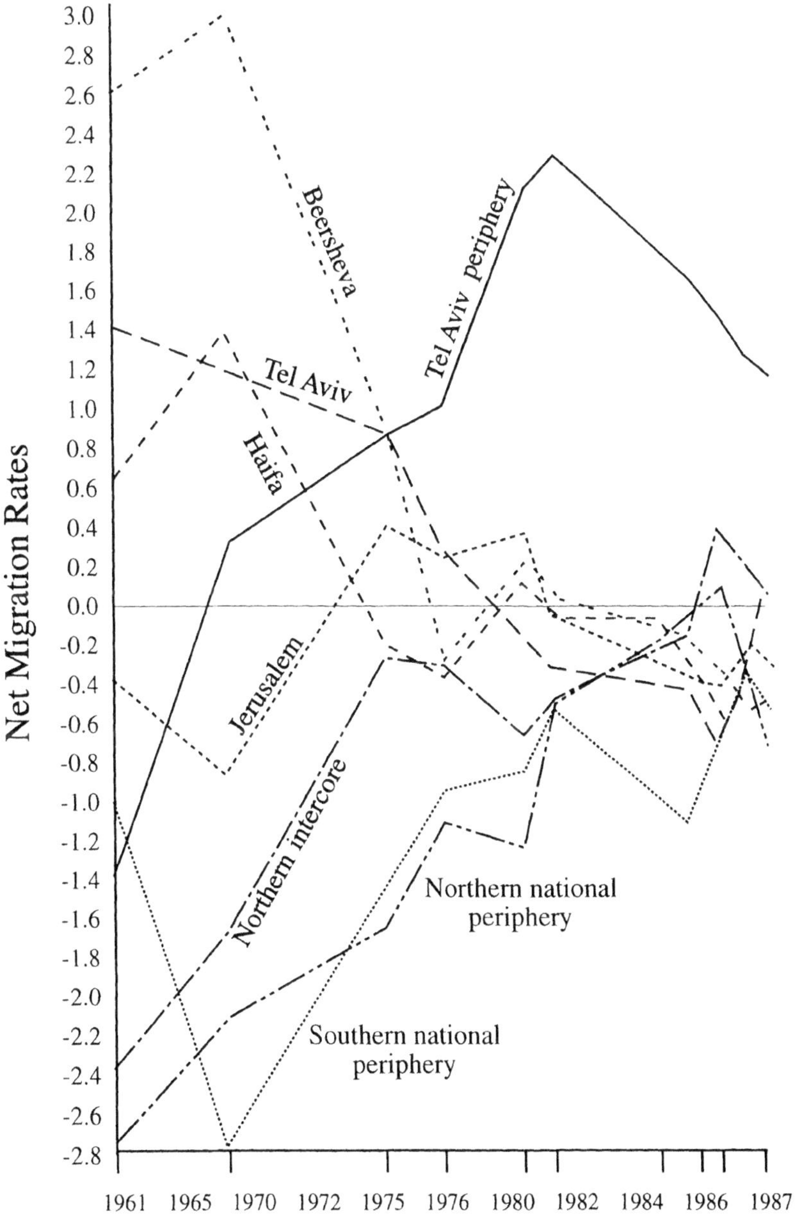

Fig. 17b: Net migration rates within the core-periphery structure, 1961–1987 (percent)

periphery has led to decentralization of the population on the one hand, and to centralization of the area of employment on the other. This process has increased the volume of commuting from periphery to center, involving a more time-consuming "journey to work." Families in their fifties who moved to the suburbs in their thirties make up the majority of those returning to the city of Tel Aviv. Now that their children have grown up and moved out (and sometimes married), they leave their large detached houses in the suburbs for high-quality, multistory buildings in the city itself. This, in addition to the gentrification process and the official municipal policy of encouraging in-town residence, has led to the positive swing.

The Haifa and Jerusalem regions, which make up the secondary core of Israel, displayed migration patterns similar to one another, but brought about by quite different factors. The Haifa region displayed a positive migration balance in the early 1960s and a negative migration balance beginning in the early 1970s. The Jerusalem region, where the level of development was low until the late 1960s, had a negative migration balance until it became positive in the mid-1970s as a direct result of the diversion of national resources to its development as the capital of Israel. In the mid-1970s, net migration in the Jerusalem region once again became negative.

In Haifa, the spatial pattern of the migration balance and the causes of this pattern are quite similar to those described above for the Tel Aviv region. However, a comparative analysis of these two regions supports the assumption of a "metropolitan maturity" factor as a catalyst in the transition of net migration from positive to negative. The Tel Aviv region, which became a metropolitan center earlier than Haifa, also began to suffer earlier from problems of what is commonly called the diseconomies of scale.

The temporal variability of net migration in the Jerusalem region differs from that in the Tel Aviv and Haifa regions, because of the socioeconomic character of Jerusalem and the official development policy concerning Jerusalem and its vicinity. Net migration in Jerusalem was negative at the start of the 1960s, but not for the usual reasons given for such a trend in a central metropolitan city. In the 1960s Jerusalem was not yet a metropolis; this was why many people left. Jerusalem of the 1960s was a border town (between Israel and Jordan), a city lacking a varied employment base and with little diversity in culture and entertainment. Entrepreneurs and businesspeople were forced to realize their aspirations elsewhere, usually in the Tel Aviv metropolitan area. Most jobs in Jerusalem were in the public services: government, municipality, university, hospitals, Jewish Agency, courts, and other public institutions. In short, Jerusalem of the 1950s and 1960s was an unattractive place. In 1965, for instance, 2,253 persons left the Jerusalem region for the Tel Aviv region, representing 33 percent of all those leaving the Jerusalem region. By the end of the 1960s, 40 percent of out-migrants from the Jerusalem region were bound for the Tel Aviv region.

The net migration of the Jerusalem region turned positive early in the 1970s, after the Six-Day War (1967) and after the government decided to expand the city limits. Furthermore, it was decided to build external residential quarters and to subsidize those moving to them; to place emphasis on the city's development, particularly through the establishment of high-technology industries and the development of tourism; and to transfer to Jerusalem government offices hitherto located in Tel Aviv. The building of five new neighborhoods in Jerusalem, mostly by Labour-Alignment governments (Ramot in the northwest, Neve Ya'aqov and Pisgat Ze'ev in the north, East Talpiot in the east, and Gilo in the south) was an extremely significant factor in the demographic growth of Jerusalem and the attraction of population to the city. Today each of these neighborhoods has between 30,000 and 50,000 inhabitants.

These activities were only some of the measures taken by the government to bolster Jerusalem as the capital of Israel. The policy bore fruit, and net migration in Jerusalem turned positive. In addition, many of the immigrants who came to Israel in the 1970s—most of them from North America and Western Europe—chose the Jerusalem area as their initial place of residence. The shift in net migration from positive to negative in the early 1980s was also the direct outcome of official policy. This policy, particularly from 1977 to 1984 (when the Likud, a right-wing party, was in power), was the building of urban and community settlements around Jerusalem and involved construction solely in Judea and Samaria (West Bank). In hindsight, such a policy can be seen to have contributed to the increase in the number of people leaving Jerusalem and thereby to its negative net migration. For example, in 1986 approximately 35 percent of those leaving the Jerusalem District moved to community settlements in Judea-Samaria, whereas in 1988 the figure was down to 25 percent. Undoubtedly, the dramatic decline in migration from the Jerusalem District to the Tel Aviv and Central districts as a percentage of all those leaving the Jerusalem District has to do with the high rate of migration from the Jerusalem District to Judea-Samaria, often to locations very close by.

It should be noted that the Likud government—more than the Labour government—regarded the establishment of community settlements around Jerusalem as part of an overall strategy for strengthening Jerusalem.[3] These localities essentially form the "outer ring" of the Jerusalem metropolitan area, and many of their inhabitants work in Jerusalem. Many Jerusalemites—especially young families—have moved to these localities for a variety of reasons: ideological, social, and economic.

The regions of the northern and southern national periphery (Northern and Southern districts) and the eastern intercore area presented a clear and stable pattern of migration throughout the period under study. They had negative net migration with high values. It appears that the large industrial plants in the Negev (Southern District) did not play a significant role in attracting younger and skilled migrants to the region. The low diversity of large plants and the low wages paid to workers there encouraged much of the region's population to migrate to the core regions of the country.

The situation in the Galilee (Northern District) is quite different (see Kipnis 1984; Lipshitz 1994). There is no doubt that the marked positive changes in net migration in the Galilee should be attributed to the official government policy adopted in the mid-1970s of increasing the Jewish population of the Galilee. In the 1970s it became clear to the government and the Jewish Agency that the strategies of the 1950s and 1960s were not suited to planning in the 1980s. In other words, inexpensive housing projects were no longer attractive, especially not to families in high socioeconomic

3. The Likud is a "right-wing" party in the Israeli political jargon; this means that it advocates holding on to as much of the Land of Israel (including the West Bank, Gaza Strip, and Golan Heights) as possible. Labour is a "left-wing" party, as it considers the West Bank, Gaza Strip, and Golan Heights bargaining chips to be exchanged for peace with the Arabs. Both insist on keeping all of Jerusalem within Israel's borders.

groupings. Thus the government and Jewish Agency decided to establish "community settlements" of various types, the size of each settlement varying between a few dozen and several hundred families (Lipshitz 1993a; Soffer and Finkel 1988; Law-Yone and Lipshitz 1991).

In less than ten years, about 50 community settlements were established in the Galilee, most of them suburban. Each family had its own house with attached garden, and almost all such localities had basic services. High levels of threshold services were also built in the regional centers. The socioeconomic reason for establishing these localities was to attract residents with a high socioeconomic status, especially young families with higher education and academic, scientific, and technological professions (on the geopolitical and land considerations, see Soffer and Finkel 1988; Law-Yone and Lipshitz 1991). From the planners' perspective, encouraging such people to move to the periphery was a way of "planting" highly developed human capital, which would serve as an important locational factor for high-technology industries. Within ten years these localities had drawn about 4,000 young, educated families with academic, scientific, and technological professions. Their income, level of education, and motorization rate were significantly higher than the national average, and their housing density was significantly lower. The majority of the work force commutes from these localities to the Haifa metropolitan area, while others have started small businesses in the region, some of them based on high technology (Lipshitz 1993a). This settlement project changed net migration in the Galilee from negative to positive. There is no doubt that if this settlement system had not be established, net migration in the Galilee would have remained negative, as it did in the Negev.

It should be noted that the migration trends described for the Galilee and the Negev in the 1960s, 1970s, and 1980s are true particularly for the Jewish population of these regions. Migration by the Arab population of the Galilee (the Arabs in the Negev are primarily Bedouin) during those decades was mainly within the Galilee, either from one small Arab village to another or from the village core to the periphery of the village. Small numbers of Arabs have moved to Haifa (especially since the 1980s) or to the Jewish towns in the Galilee (particularly Upper Nazareth). Thus, migration by the Arab population is not *interregional* but *intraregional*, and it is therefore beyond the scope of this book (for more detail, see Lipshitz 1991b).

Judea-Samaria, known internationally as the "West Bank," is not a significant geographical unit in terms of immigration, because very few immigrants have settled there. It is significant, however, with respect to internal migration, especially from Jerusalem. Jewish migration to Judea and Samaria began immediately after the Six-Day War (1967)[4] for national-religious reasons (settlement of all parts of the Land of Israel west of the Jordan River). It then continued for social reasons (living in a place where residents care about the social and community profile of the locality in terms of educa-

4. The Six-Day War began on June 6, 1967, and ended on June 11. In the weeks prior to the war, Israel's existence was threatened by three Arab countries that hoped to wipe it out completely: Egypt to the south, Syria to the north, and Jordan (via the West Bank) to the east. By the time the war ended, Israel had achieved a phenomenal victory, capturing the Sinai Peninsula in the south (which was subsequently returned to Egypt under the peace treaty), Judea and Samaria in the center, and the Golan Heights in the north.

tion, relations between individuals and the community, and so on) and economic considerations (relatively low prices for low-rise housing). Between 1967 and the late 1980s approximately 120 localities were established in Judea and Samaria, and they drew a total of 125,000 migrants. Net Jewish migration in Judea and Samaria has been positive since 1967, because few Jews leave these regions.

For the most part, the considerations of the migrants (or "settlers," as they are more often called) to localities in western Judea and Samaria are economic and social. Most of these localities belong to the distant ring of the Tel Aviv, Jerusalem, and Haifa metropolitan areas. Many of the residents (especially the men) work in these metropolitan labor markets. In contrast, the considerations of the migrants/settlers who live in the heart of Judea and Samaria are mainly religious/ideological. Presumably, the final-status agreement between Israel and the Palestinian Authority will have an impact on the trends of migration of the Jewish population to and from Judea-Samaria.

As stated previously, immigration to Israel fell drastically in the early 1960s and remained low until the late 1980s. Consequently, net internal migration and natural increase (the balance of births and deaths) became the main factors determining population growth trends. While net internal migration was negative in the peripheral districts and positive in the Central District, natural increase in the peripheral districts exceeded that in the Central, Tel Aviv, and Haifa districts for two reasons: a high proportion of immigrants who arrived in the 1950s from North Africa and Asia; and a high percentage of Arabs.

As figure 18a shows, net migration ultimately had a major impact on geographical trends of population growth. The population of the Tel Aviv District—where net migration is negative (due to suburbanization to the Central District)—as a percentage of the total Israeli population is decreasing, and that of the Central District is increasing

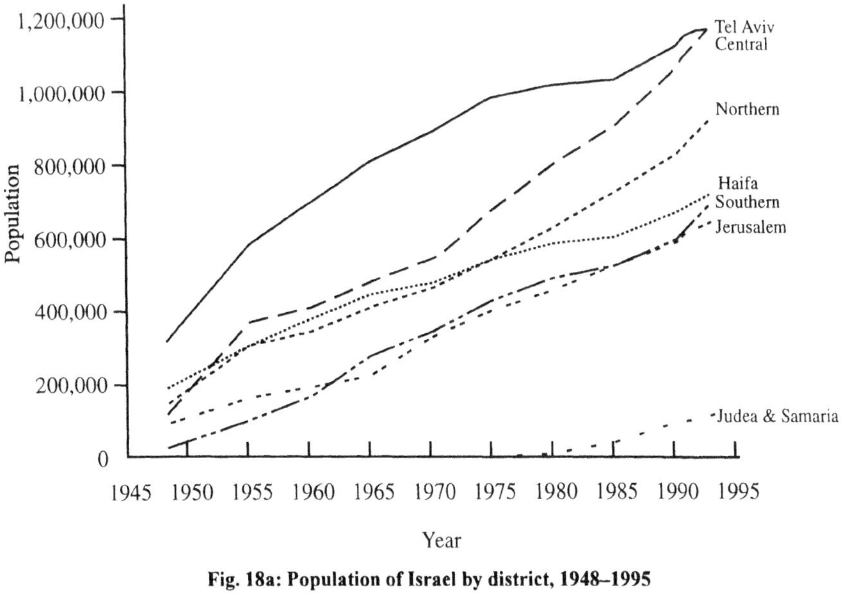

Fig. 18a: Population of Israel by district, 1948–1995

(fig. 18b). In 1990 these two districts were home to 50 percent of the total Israeli population (fig. 18b). Figure 18b also shows that the share of the population living in the peripheral districts rose substantially until the mid-1960s—because of immigration in the 1950s—and then remained almost constant. In several peripheral development towns the population did not grow between the mid-1950s and the late 1980s. In other words, the population growth in these towns due to natural increase was offset by negative net internal migration.

The share of the population in the Jerusalem District (in which the city of Jerusalem is the main locality) dropped from the establishment of the State until 1967. As described above, Jerusalem was an economically unattractive border town. In 1967 the city was reunited, and major government incentives were offered to promote its economic and physical development. As a result, the population growth rate in the Jerusalem District rose beginning in 1967, and population of the district as a percentage of the total Israeli population also increased somewhat (figs. 18a and 18b). Beginning in the late 1970s, this trend was halted in the Jerusalem District due to the establishment of numerous localities in Judea and Samaria (especially during the tenure of the Likud government) and migration from Jerusalem to these localities. It seems that the growth in the Jewish population of Judea and Samaria and the increase in the share of this region in the total Israeli Jewish population explains why the proportion of the Israeli Jewish population living in the Jerusalem District stopped growing, and the rate of population growth in the Central District decreased slightly (figs. 18a and 18b). Long-range suburbanization to Judea and Samaria affects the population growth rates of the Jerusalem and Central districts.

Since 1990, the data presented in figures 18a and 18b have changed somewhat due to immigration from the former Soviet Union, as discussed in chapter 4.

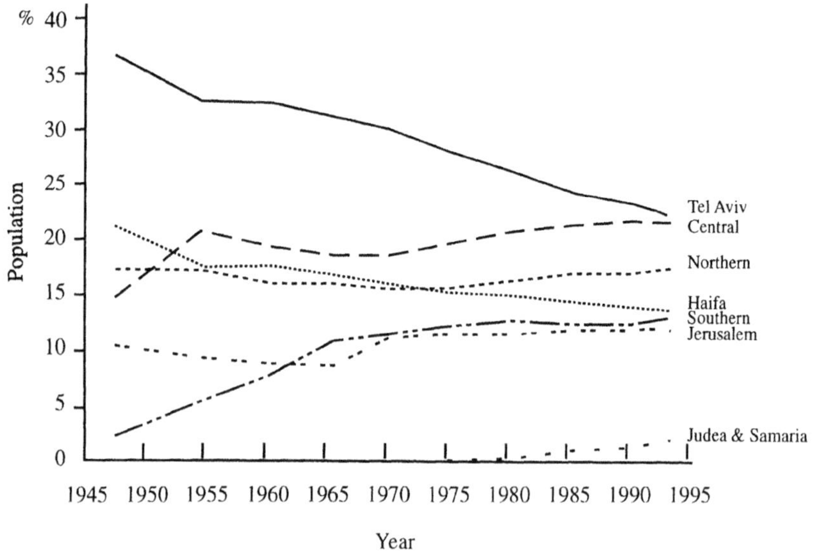

Fig. 18b: Population of each Israeli district as a percentage of the total Israeli population, 1948–1995

3.5 The Preferences of the Migrants within the Core-Periphery Structure

The spatial distribution of regional net migration is the result of the spatial distribution of migrants' preferences. The spatial preferences of internal migration in only three regions will be discussed here: the Northern District (northern national periphery), the Southern District (southern national periphery), and the Tel Aviv core area.

During the period under study (1961–1988), significant changes took place in the spatial results of migration in the Southern District. Figures 19–22 show that from 1961 to 1965 the percentage of migrants from the Tel Aviv, Jerusalem, and Haifa districts to the Southern District increased. (The percentage of migrants from the area of origin to the destination area is the percentage of total migration from the area of origin.) For example, in 1961 the percentage of migrants to the Southern District from the Tel Aviv District was 18 percent, from the Jerusalem district 12 percent, and from the Haifa district 10 percent, while in 1965 the percentages rose to 31, 32, and 22 percent, respectively (figs. 19–22).

Although the percentage of migrants from the Southern District to metropolitan areas rose in the same period, the new result favored the Southern District. In 1965, for example, 2,304 people moved from the Jerusalem District to the Southern District, while only 295 moved in the opposite direction. The fact that in 1965 net migration in the Beersheva area (the "capital" of the Southern District) compared favorably to all other regions of the country also indicates that the south was the main target of internal migration in Israel. In addition, it is noteworthy that from 1961 to 1965 only two regions in Israel had a positive migration balance: the Tel Aviv District and the South-

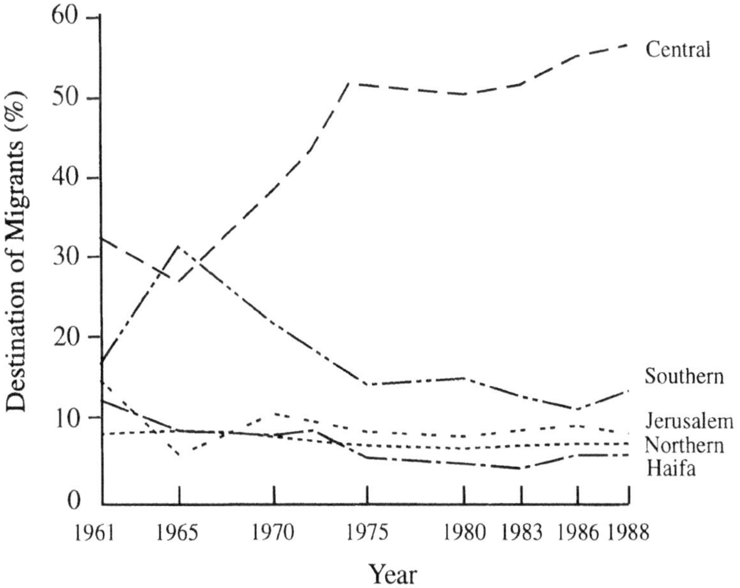

Fig. 19: The distribution of migration rates from the Tel Aviv District to other districts, 1961–1988

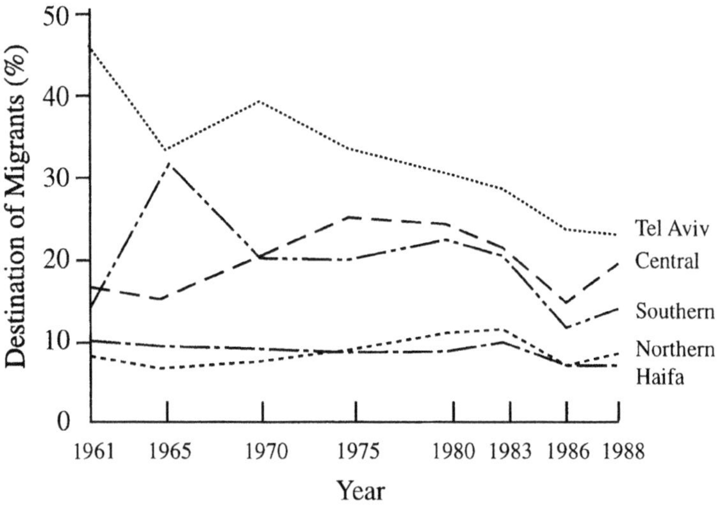

Fig. 20: The distribution of migration rates from the Jerusalem District to other districts, 1961–1988

ern District. It is clear, therefore, that in the early 1960s internal migration in Israel brought about the dispersal of the population to the Southern District, especially the Beersheva region (this occurred in the early 1960s only, and was reversed in the late 1960s). This spatial result clearly reflects the Israeli government's policy of encouraging migration from the center of the country to the national periphery (northern or southern) during this period. This encouragement included loans to private enterprise, substantial subsidies to apartment purchasers, provision of infrastructure for factories, tax reductions, and so on. Was the process of population dispersion to the Southern District also accompanied by a dispersion of development?

From the few existing statistics it appears that internal migration in the early 1960s also served to direct development to the Southern District and added an above-average population to the area. For example, in 1961 net migration of residents with at least 13 years of education was +1,020 in the Southern District, +520 in the Tel Aviv District, and +740 in the Jerusalem District (=16.9 percent of the population with 13 or more years of education in the Southern District, 0.94 percent in the Tel Aviv District, and 3.43 percent in the Jerusalem District), while the figures in the other districts were negative (Lipshitz 1991).

Net migration among two ethnic groups, native-born and European- and American-born (which represented a higher socioeconomic level within Israeli society, especially in the 1950s and 1960s), was as follows: The migration balance of native-born Israelis in 1961 was +3,715 in the Southern District and +6,725 in the Tel Aviv District (=6.3 percent of the native-born population in the Southern District and 2.6 percent in the Tel Aviv District), while in the other parts of the country it was negative. The migration balance of European- and American-born individuals was +1,295 in the Southern District and +13,000 in the Tel Aviv District (=2.86 percent of European- and American-born population in the Southern District and 5.09 percent in the Tel Aviv

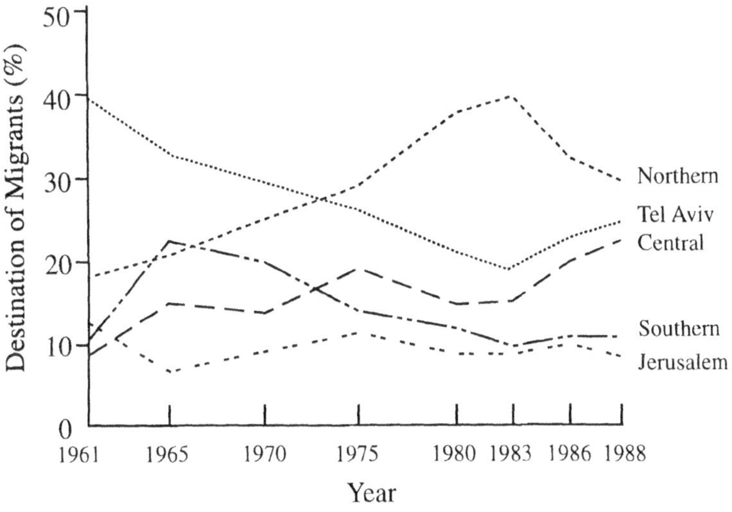

Fig. 21: The distribution of migration rates from the Haifa District to other districts, 1961–1988

District), while in the rest of the country it was negative. It should be noted that in the 1950s and 1960s there was a significant and positive correlation between the geographical distribution of native-born and European- and American-born populations and the geographical distribution of income, education, and other variables. The composition of the migration to the Southern District in the early 1960s created a strong potential for growth.

In the mid-1960s a dramatic decrease took place in the percentage of migrants from the Tel Aviv, Jerusalem, and Haifa districts to the Southern District (figs. 19–22); at the same time there was a rise in the rate of migration from the Southern District, especially from Beersheva to the Tel Aviv area and the Central District. This process caused net migration in the Southern District to decline, becoming negative in 1970 (see figs. 17a–17b). The development potential of the migrants from the Southern District was high, and thus their out-migration was detrimental to the socioeconomic base of that district (Lipshitz 1987).

In the mid-1970s, migration from the Tel Aviv and Central districts to Judea and Samaria, especially the western fringes of Samaria, gathered speed. For example, approximately 10 percent of those leaving the Tel Aviv District in 1986 and 1988 chose to move to Judea and Samaria. In effect, this process caused the expansion of the Tel Aviv metropolitan area, as a high proportion of migrants who left the Tel Aviv and Central districts for Judea and Samaria remain in the labor market of the Tel Aviv metropolitan area.

Quite likely, the government's investment in the educational system in the national periphery resulted in an increase in the percentage of the population from the national periphery continuing their education in institutions of higher learning in the center of the country. However, the absence of work in their regions of origin that was suited to the skills they had acquired at the universities in the center of the country meant that most university graduates remained in the central region. The result was a polarization of both population and development.

The spatial results of migration in the Northern District (northern national periphery) are different from those noted for the southern national periphery. Figures 19 and 20 show that, during the period when the migration rate from the Tel Aviv and

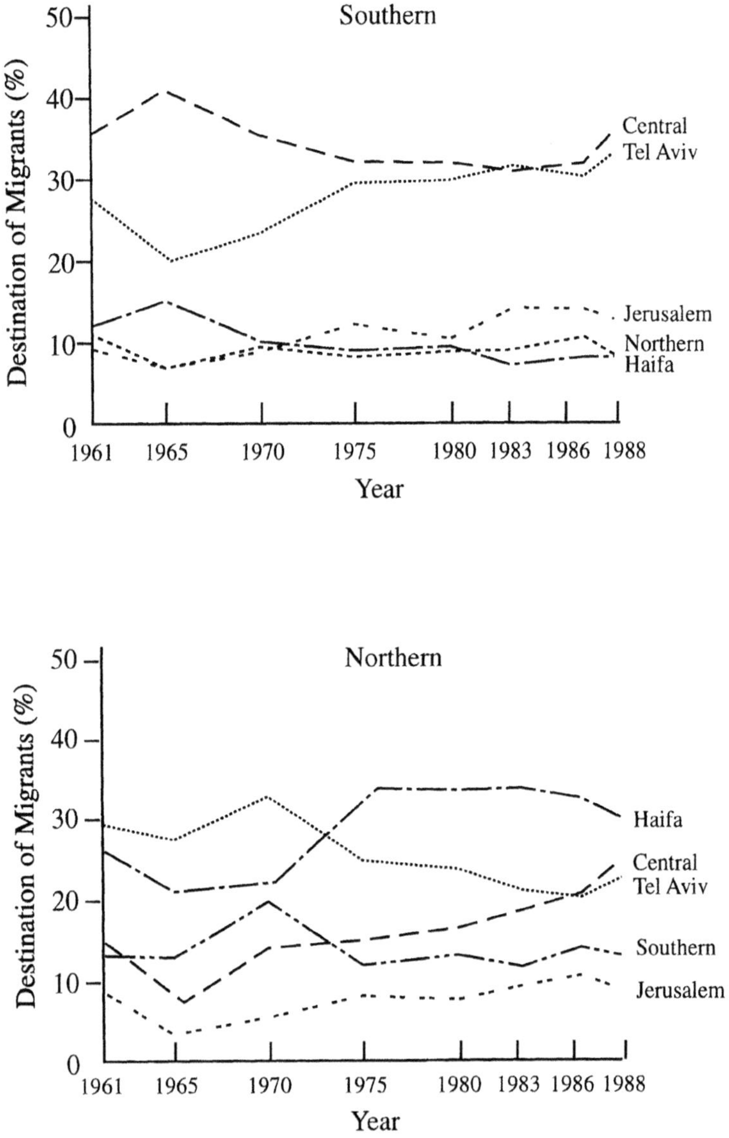

Fig. 22: The distribution of migration rates from the Southern and Northern districts to other districts, 1961–1988

Jerusalem districts to the Southern District increased, the percentage of migrants from these districts to the Northern District did not increase and remained stable at low rates. During the same period (especially in the 1960s), the percentage of migrants from the Northern District to the Central District rose (fig. 22). In absolute terms it appears that more people migrated from the Northern District to the core districts (Tel Aviv, Haifa, Jerusalem, and Central) than in the opposite direction, and the Northern District experienced negative net migration during the 1960s and 1970s (see figs. 17a–17b).

From the little statistical data available, it appears that the negative net migration reflects the departure of the above-average socioeconomic population from the Northern District. For example, in 1961 the migration balance of people with 13 or more years of education in the Northern District was –965, and the migration balance of those with 11–12 years of education was –750; the migration balance of native-born Israelis was –4,095, and that of European- and American-born Israelis was –4,600. This brain drain at the beginning of the 1960s continued into the 1970s (Lipshitz 1987). In summary, internal migration in the Northern District acted as a polarizing factor not only by reducing the number of inhabitants but also by reducing the socioeconomic level of the area.

In the mid-1970s a reversal took place in the migration balance of the Northern District (see figs. 17a–17b). Beginning in 1975, net migration in the north became less negative, and at the beginning of the 1980s it turned positive. One can conclude from figure 7 that this change was due mainly to the substantial rise in the percentage of migrants from the Haifa District to the Northern District. In 1983, approximately 40 percent of all those who left Haifa migrated to the Northern District, while in 1965 the figure was only 20 percent. A detailed analysis of the process seems to indicate that most of the migrants from the Haifa District chose one of two regions as their destination: the western Lower Galilee or the western Upper Galilee (Sonis 1988; Lipshitz 1991a, 1991b). In the last decade, most of the migrants from Haifa to these areas chose the newly conceived community settlements as their destination. These settlements were established either directly by the Israeli government and Jewish Agency or with indirect government assistance. Each of these communities had between 50 and 100 families, each with a private home and garden. The socioeconomic level, quality of life, and cultural activity in these localities are of a high standard. The population migrating to the community settlements is characterized by a high socioeconomic level. About 80 percent have 13 years or more of schooling (the national average is approximately 25 percent). Some of the migrants continue to work in the Haifa District; others work in the science and high-technology industries that were established in these areas on the basis of the high quality of human resources that migrated to the area in the mid-1970s. It appears, therefore, that the population of high socioeconomic standing that migrated to the Northern District chose as its destination the areas relatively close to Haifa, and that this migration is actually part of the long-distance suburbanization process taking place in metropolitan Haifa.

The dispersion of population and development on the metropolitan level, as it occurred in the Haifa area in the 1980s, had already begun in metropolitan Tel Aviv (which is more mature than metropolitan Haifa) in the mid-1960s (Shachar 1975; Gonen 1978). Calculations show that since the mid-1960s the migration rate from the city of Tel Aviv to its periphery has been on the rise. In 1983 about 50 percent of those leaving Tel Aviv migrated to its periphery, compared to only 18 percent in 1961. Figures 17a and 17b show that a rise in the positive migration balance in the Tel Aviv periphery is accompanied by a rise in the negative migration balance in the Tel Aviv District. Calculations show that from 1961 to 1988 the rate of migration from the Tel Aviv District to each of the regions on its periphery rose consistently.

In the 1960s the destination preferred by those leaving the Tel Aviv District was Petah Tiqva, which constitutes the eastern periphery of Tel Aviv. In the 1970s the Petah Tiqva region became less attractive, and the Rishon Lezion region, situated on the southern periphery of Tel Aviv, then became the main destination for out-migrants from the Tel Aviv region. In the 1980s the main target was the southern Sharon region, situated on the northeastern periphery of the Tel Aviv area. This demonstrates that the geographical direction of population dispersal and development in metropolitan areas varies with time. It can be assumed that the change in geographical direction of population dispersal, as presented above, is in fact the spatial expression of private enterprise considerations (the supply side) and of consumer preferences (the demand side). In other words, developers and contractors seek available (and relatively inexpensive) land for building. When a particular area is saturated with housing and land prices rise, they move to areas where land for building is more plentiful, and they develop residential neighborhoods there.

Research conducted in the 1970s showed that the migration absorbed by the Tel Aviv District lowered the Tel Aviv metropolitan averages for income and education (Shachar 1974; Gonen 1978; Shachar and Lipshitz 1980). Meanwhile, the percentage of adults and senior citizens in the Tel Aviv core has continued to increase compared to the average age distribution of the Tel Aviv periphery. In contrast, migrants to the Tel Aviv periphery are generally young families with high levels of education and income. Shachar points out that by the mid-1960s people born in Israel, Europe, and America were already conspicuous within the group leaving the Tel Aviv District (Shachar 1974). As a result, net migration among this group was positive in the Tel Aviv periphery and negative in the Tel Aviv core. As stated earlier, the geographical distribution of Israel's population in the 1950s and 1960s by country of origin had a clear economic and social significance. Migration from the Tel Aviv core to the Tel Aviv periphery generated decentralization of well-being and development from the primary core of the country to the periphery of that core.

We can therefore conclude that in metropolitan Tel Aviv the spatial result of population dispersion and development dispersion on the metropolitan level became apparent ten years earlier than in metropolitan Haifa. Meanwhile, the process of population dispersion and development dispersion began in metropolitan Jerusalem (which is less mature than Tel Aviv and Haifa in terms of metropolitan development) only in the early 1970s (Lipshitz 1987).

Migration from the Jerusalem District to its periphery was mainly directed toward the small towns and community settlements located within 30 minutes' commuting distance from Jerusalem. These areas were established mainly by the Israeli government and slightly by private enterprise. More than 90 percent of the Jewish localities in Judea and Samaria were established between the mid-1970s and the late 1980s, and the large-scale migration to the community settlements around Jerusalem took place during that period. The few surveys conducted in these localities show that the migrant population has a very high socioeconomic level. In other words, the newly founded settlements around Jerusalem drew highly developed, high-caliber human capital from Jerusalem. The impact on the economic development of Jerusalem has been marginal, however, because the number of migrants is relatively low (compared to the population of Jerusalem), the quality of human capital that remained in Jerusalem is high, and most of the migrants continue to work in Jerusalem.

To sum up, internal migration in Israel, as elsewhere in the world, is from the peripheral regions to the core. In the 1960s it was mainly to the core regions of Tel Aviv and Haifa, whereas in the 1970s and 1980s it was mainly in the direction of the periphery of the Tel Aviv core (the Central District). The Central District is becoming a destination not only for migrants from the peripheral regions, but also for those leaving the Tel Aviv core (the city of Tel Aviv and the inner ring of the metropolitan area). Those moving out of the peripheral districts are usually young, educated people with academic and scientific professions. This migration harms the development potential of the peripheral regions and contributes, on a cumulative basis, to the economic development of the core regions.

Internal migration in Israel acts on two geographical levels: the national level and the metropolitan level. On the national level it increases core-periphery polarization, and on the metropolitan level it leads to decentralization of the population and to a reduction in the economic and social gap between the Tel Aviv and Haifa cores and their peripheries (urban and rural suburbs).

CHAPTER 4

THE GEOGRAPHICAL DISTRIBUTION OF THE IMMIGRANTS FROM THE FORMER SOVIET UNION IN THE 1990s

4.1 The Spatial-Policy Dilemma in the Housing Market: The National and Regional Levels

As we saw in previous chapters, there was little immigration to Israel from the early 1960s until 1989. During this period, 10,000–15,000 immigrants arrived each year, most of them from North America and Western Europe. Many of them were religiously motivated, and therefore a large percentage chose to settle in Jerusalem and Judea-Samaria. A large portion of the newcomers were fairly affluent and well educated, and they settled in the major cities and their environs.

In late 1989, and even more so in 1990, a new wave of immigration began. This wave peaked in 1990, when 200,000 immigrants arrived in Israel. Between 1990 and 1995, 700,000 immigrants arrived (see fig. 23), most of them from the former Soviet Union. Since 1992, between 60,000 and 70,000 immigrants a year have been coming from the former Soviet Union, an average of 5,000–6,000 a month (see fig. 23).

The mass immigration from the former Soviet Union created an immediate need

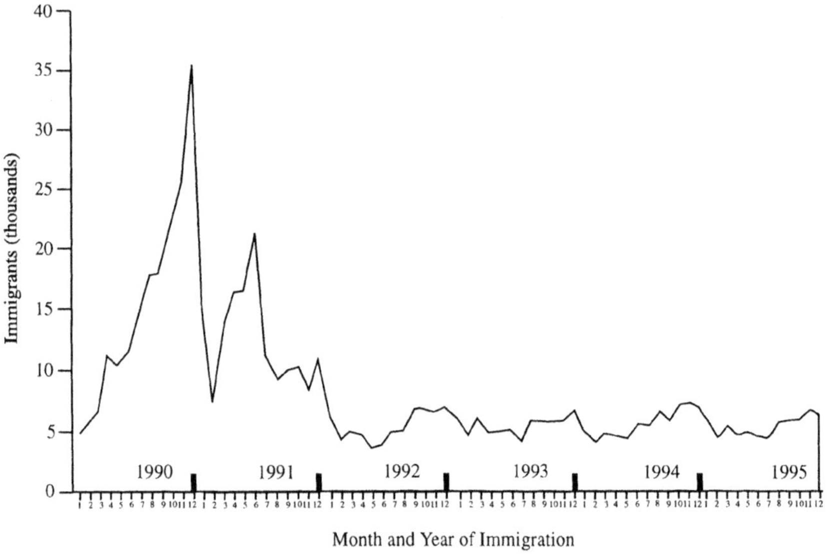

Fig. 23: Immigrants from the former Soviet Union, by month of immigration, 1990–1995 (source: Central Bureau of Statistics, *Immigration to Israel 1995*)

for housing for thousands of families. Roughly 145,000 apartments were needed in the first two years alone, and 215,000 in 1990–1994. The gap between supply and demand was particularly wide because in the years preceding the wave of immigration, the volume of building in Israel had dropped to 20,000 dwelling units a year (Borochov 1991).

One of the first and most important dilemmas that faced Israeli policymakers when hundreds of thousands of immigrants began to arrive was whether, in the age of accelerated development spurred by immigration, government intervention should be used to influence the geographic dispersion of immigrants or whether market forces should be given free rein. Should the Ministry of Construction and Housing intervene in the spatial distribution of housing (the number of dwelling units in each region)? Should the Ministry of Industry and Trade intervene in creating jobs in the periphery? Or should market forces be allowed to determine the distribution?

In the first year of the wave of immigration, the government adopted a strategy of *absorption by market forces*, which is essentially the *"direct-absorption policy*." This policy constitutes a reversal in the manner of thinking regarding immigrant absorption. Upon landing in Israel, immigrants now receive an "absorption package" from the government, containing a certain sum of money—roughly $10,000 for a family of four—meant to last for one year. With this sum they may do as they see fit in all areas of life: housing, employment, education, consumption, etc. Direct absorption thus lets immigrants decide for themselves, based on their own personal considerations, how they wish to be absorbed. The declared goal of the government's direct-absorption policy was to lessen immigrants' dependence on national and local government; to streamline the absorption process itself and make it more flexible; and to speed up the immigrants' ability to start functioning in housing, employment, and social and cultural life. In other words, the Israeli government decided *to replace its strategy of intervention from above with a strategy of absorption by market forces.*

Spatially, this policy was "pure" only until the end of 1990, when *"intervention from above"* began in the housing market. Until then the Israeli government had believed that the regular housing market would suffice, and there would be no need for special government intervention. It was assumed that 20,000–40,000 immigrants would arrive each year, and this many people could be absorbed by the housing and job markets with no major shocks. However, when the government realized in late 1990 that it should expect 200,000 immigrants per year, and one million immigrants in 1990–1995, it changed its spatial housing policy and decided to intervene directly in the number of dwelling units to be built and their geographical distribution. Its approach was still different from that adopted in the 1950s, though. In the 1950s the government not only built homes in the peripheral districts; it also initially directed the immigrants to them (in many cases immigrants were taken from the ship or airplane straight to apartments in peripheral development towns). In the 1990s, the government decided to build homes on a large scale in the peripheral districts, but let the individual immigrants choose where to live. Massive government building was carried out in the Northern and Southern districts because of State-owned land reserves in these districts, as we shall see in more detail below. Because the central government owned extensive land resources, it was able, in conjunction with the local authorities, to develop new neighborhoods, including housing, commerce, public services, and open spaces.

Given that the government would intervene, it was faced with the following question: should it encourage immigrants to settle in the peripheral regions of Israel or should it create a high-quality housing market in the periphery that would attract nonimmigrants and well-to-do immigrants, leaving homes available in the central region for the immigrants via a trickle-down process? If the latter approach were to work, it would raise the socioeconomic level of the periphery and create a higher profile of human capital there that would form the basis for the development of a modern economy.

There were two main approaches to this question. The first argued that the work of population dispersion should be left to the nonimmigrants, while the immigrants find jobs and assimilate into social networks in central Israel (Gonen 1991). This argument relies on the slow spatial processes of long-distance suburbanization among nonimmigrants (some of whom would continue to commute to the metropolitan areas) that have taken place in Israel in the past decade and on the good job prospects for the immigrants in the metropolitan areas (where they cannot work if they live in the periphery).

Underlying this approach was the assumption that the nonimmigrant population, which in the previous decade had begun to disperse to community settlements in the Galilee and the Negev and to the "suburban" neighborhoods of development towns in these regions, would bring financial capital and high-quality human capital to the peripheral regions. The country would not have to expend extensive national resources on helping the nonimmigrants resettle. According to Gonen (1991), a solid, middle-class population could settle in the periphery and thereby achieve the goals of the dispersion policy without being nearly as much of a drain on national resources as immigrants who lacked the traits of the established population would be. Their steady, relatively high income and high motorization rate would enable a portion of this population group to have a high quality of life in the national periphery and work in the metropolitan areas and their peripheries. Moreover, this dispersion would give the national periphery the important locational factor for the development of high-tech industries—human capital. According to this approach, encouraging nonimmigrants to move to the peripheral regions—especially to high-quality homes in a superb social environment—would leave apartments in central Israel available for the immigrants. Moving immigrants to the peripheral regions, however, would require substantial national resources to build apartments and develop jobs suited to the immigrants' skills. Settling tens of thousands of immigrants in the peripheral regions without creating suitable jobs might quickly lead to the migration of "strong immigrants" to the central region. The result might be not only a short-term stock of empty housing, but also a deterioration of the economic development potential of the periphery.

This approach also uses a sociospatial argument. Most long-time residents of the peripheral regions are themselves first- and second-generation Israelis from Asia and North Africa. Over the years, this population group has experienced social, economic, and political mobility. "Sending large numbers of Soviet immigrants—and most of these immigrants are European—[to the peripheral regions] would introduce competition for jobs and political positions into the peripheral regions" (Gonen 1991, p. 35). Such competition could engender ethnic social tension and waste the local and regional growth potential inherent in the immigration.

The second approach argues that the long-term suburbanization process is slow, and although it picked up speed in the early 1980s, it was slowing down again by the end of the decade. If we truly want to develop the peripheral regions and bring some of the towns to the "takeoff point," then we have to take advantage of the waves of immigration, because the mechanism of internal migration has not produced critical changes in the peripheral towns and is unlikely to do so in the short or medium term. The present immigration, argue the advocates of this approach, differs from that of the 1950s in that the newcomers in the peripheral regions are educated, and a high proportion of them have academic and scientific professions. Suitable jobs must be developed in the periphery so that these people do not leave for central Israel; such jobs would boost the development potential of the peripheral regions (Gradus 1990; Kipnis 1990).

Some of the advocates of this approach believe that the government should intervene to create development centers in the peripheral regions, too. In other words, the development of housing and jobs should focus on two centers: Beersheva in the Negev, and a triangle bounded by lines connecting Misgav, Karmiel, and Ma'alot in the Galilee (Kipnis 1990). According to this approach, a growth center around Beersheva would produce a "threshold" for high-quality services on a large scale. These services would enhance the quality of life and support not only the residents of the city but the inhabitants of the entire region (Krakover and Stern 1990).

National Outline Plan (NOP) 31, an integrated plan for construction, development, and immigrant absorption, in a certain sense combines the two approaches described here: accelerated planning and development in central Israel and accelerated development in two peripheral centers. The plan gives top priority to bolstering the status of the four metropolitan areas (while establishing the fourth—Beersheva). According to the plan, development of the four metropolitan areas must be combined with the following activities: strengthening the ties between the central region and the periphery; improving access of residents of the periphery to jobs and services in the central region; and extending the commuting field. Thus, NOP 31 does not agree with the approach that advocates "taking advantage" of the immigrants for the purpose of accelerated development of the periphery. It calls for spatial development in keeping with the trends produced by market forces, rather than fighting these trends. In other words, development of the core regions should be accelerated, and another core in the south around the city of Beersheva should be established. The plan assumes that this will cause the availability of housing and jobs to trickle down to regions located on the "seam" between the primary core and the national periphery (Lehrman et al. 1992).

Another approach worth noting here advocates government intervention in population dispersion, including the dispersion of the immigrant population, but places particular emphasis on government intervention in investments in *transportation infrastructure* (Sheinin 1991). Investments in transportation infrastructure, together with subsidies for high-quality housing and tax benefits for all production factors (not only for capital), would shorten the commute from the periphery to the metropolitan areas, thereby increasing the feasibility of living in the periphery and working in the central region. This would give people (immigrants and nonimmigrants alike) three options: inexpensive, high-quality housing in the periphery and a reasonable travel time (substantially less than today) to jobs in metropolitan areas; inexpensive, high-quality housing in the periphery and jobs in the periphery; or relatively expensive housing in the central region and jobs in the central region.

4.2 The Supply Side: Spatial Distribution of Public and Private Residential Building

Some of the dilemmas presented above were relevant in the first few months after the immigrants began to arrive, when the annual rate of immigration was in the tens of thousands. When the government realized, however, that 200,000 immigrants per year (about one million in five years) could be expected, it started to intervene vigorously in the housing market. At the same time the private sector was also active in this market.

Table 4 and figures 24 and 25 show clearly that, while the public sector built more in the periphery—where the bulk of public land reserves were located—the private sector built more in the central region. In 1992, for example, the government completed 4,500 dwelling units in the Central District and 20,000 in the Southern District; in the same year the private sector completed 5,800 dwelling units in the Central District and only 3,000 in the Southern District. In 1993 the private sector completed 1,800 dwelling units in the Southern District and 6,500 in the Central District, while the public sector completed 7,800 in the Southern District and only 1,800 in the Central District. In 1990–1994 the public sector built 12,000 dwelling units in the Central District and 40,000 in the Southern District; during the same period, the private sector built 28,000 dwelling units in the Central District and 10,000 in the Southern District. In the Tel Aviv District, the public sector builds very little compared to the private sector. The ratio of apartments built by the public sector in the Tel Aviv District to those built by the private sector is 1 to 9, and in the Central District it is 1 to 2.5 (table 4). In the periphery the ratio is reversed: between 4 and 7 to 1 in the Southern District and 1.5 to 1 in the Northern District.

Table 4 also shows a clear government preference for building in the Southern District over building in the Northern District. More dwelling units are built each year by public companies in the Southern District than in the Northern District. These disparities are more significant the farther one moves from the peaks of 1992. The ratio of publicly built dwelling units to privately built dwelling units in 1990–1994 is between 4 and 7 to 1 in the Southern District, but close to 1 in the Northern District. A closer look at Housing Ministry publications indicates that public building in the Southern District was concentrated in four towns: Qiryat Gat, Ashdod, and Ashqelon (all of which are within commuting range of Tel Aviv), and Beersheva. In other words, *public building in the Southern District is concentrated in cities located within the spillover range of the Tel Aviv metropolitan area and in the fourth-largest metropolitan center: the city of Beersheva.* The policy of the Ministry of Construction and Housing is clear: instead of carrying out massive residential building in the remote periphery, where jobs require geographic proximity to one's home—i.e., there is a "*closed spatial economy*"—it would concentrate its building in development towns located within the commuting range from metropolitan employment centers (although a bit farther than the average Israeli commute). The new immigrants—like the nonimmigrants—who choose to live

Table 4: Housing Completions by District, 1990–1994
(percent of dwelling units completed)

Year	Developer	Total (absolute numbers)	Total	Jeru-salem	Northern	Haifa	Central	Tel Aviv	Southern	Judea-Samaria
1990	Private	16,710	100.0	8.3	24.5	10.7	24.8	18.9	9.9	3.9
	Public	3,250	100.0	15.1	4.9	11.4	22.2	5.2	20.3	20.9
	Total	19,960	100.0	9.4	21.4	10.7	24.4	16.7	10.7	6.7
1991	Private	19,630	100.0	6.3	21.1	9.8	29.3	16.9	11.1	5.5
	Public	23,020	100.0	4.7	24.9	10.9	11.9	1.1	41.7	4.8
	Total	42,650	100.0	5.4	23.2	10.4	19.9	8.4	27.6	5.1
1992	Private	21,590	100.0	5.5	21.0	10.1	26.6	18.1	14.2	4.5
	Public	48,530	100.0	5.7	19.9	6.8	11.5	1.5	42.5	12.1
	Total	70,120	100.0	5.7	20.2	7.7	16.2	6.6	33.8	9.8
1993	Private	21,580	100.0	6.1	22.0	11.1	29.6	17.7	8.8	4.7
	Public	21,440	100.0	12.5	11.3	13.3	8.4	2.8	36.7	15.0
	Total	43,020	100.0	9.2	16.9	12.2	19.0	10.3	22.6	9.8
1994*	Private	24,480	100.0	5.8	15.7	10.9	37.4	15.8	11.1	3.3
	Public	9,450	100.0	16.0	10.7	16.4	22.6	3.5	21.5	9.3
	Total	33,930	100.0	8.6	14.3	12.5	33.3	12.3	14.0	5.0

*January-September only.
Source: Central Bureau of Statistics, *Construction in Israel*, 1990–1992, 1994.

in these towns see an "*open spatial economy*" along the axis of place of residence (the development town) and place of work (the metropolitan area). Another trend that can be seen from figures 24 and 25 is the rapid reaction of the public housing market to the increasing demand for housing. The moment demand drops, supply drops accordingly. Therefore the supply of homes rose sharply between 1990 and 1992 and dropped sharply from the 1992 peak to 1994. There is no doubt that this adjustment—the sharp change—had a profound effect on the building industry and made orderly planning of construction difficult. These sharp changes are not seen in the private market, which apparently responded to real demand rather than expected demand; the former depends on the number of immigrants who arrive each month and each year.

To solve part of the housing problem, the Israeli government encouraged private owners to rent out apartments. Although there were plenty of empty, publicly owned apartments in the development towns (especially in the "weak neighborhoods") because of years of out-migration from these towns, the Israeli government took advantage of this available housing only when immigration increased beyond expectations. When it became clear that even the supply of empty apartments in the development towns would run out quickly, residential building began on an unprecedented scale in the development towns in the Galilee and the Negev (for an evaluation of this policy, see Shefer 1991).

The massive government building in the development towns of the Galilee and Negev ran counter to the direct-absorption strategy in the sense that direct absorption—whether planned this way or not—increased the likelihood of stimulating the housing market in the central region and perpetuating the relative freeze in the development towns. The direct-absorption strategy essentially confronted the immigrants with a tradeoff: employment or housing. If they agreed to compromise on housing and live in crowded conditions in the central region—sometimes with several families sharing an apartment—paying high rents or high purchase prices, they would improve their job prospects; but if they chose to pay low rent or low purchase prices for relatively large, spacious apartments in the development towns of the Galilee and the Negev, they would reduce their chances of finding jobs that suit their occupations and level of education.

Housing hundreds of immigrants in empty apartments in development towns and building hundreds of new apartments in these towns necessarily has social, community, and physical effects and ramifications (Bar 1990). New housing built for immigrants and young couples alike could serve, under certain conditions, as a positive turning point and, under other conditions, as a negative turning point in the evolution of the social and physical structure of the development towns in the Israeli periphery. The kind of planning done, the location of the new neighborhoods, the methods used to populate old and new neighborhoods, the type of housing, and housing prices all had a considerable effect on the physical and social structure of the development towns. The Build-Your-Own-Home project and low-rise buildings with four- or five-room apartments almost guarantee that the residents of these homes will come from the middle and upper socioeconomic classes. On the other hand, high-density construction of two- or three-room apartments dictates that lower-class families will move in (Bar 1990).

The government intervened in the housing market in three ways: First, it gave landlords tax exemptions (even for renting out more than one apartment). This mainly freed up apartments in central Israel and the big cities. Second, it permitted expedited

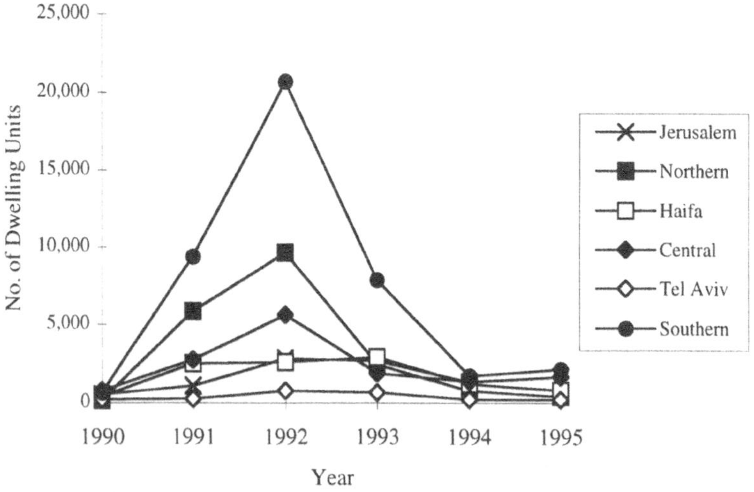

Fig. 24: Number of dwelling units built by the public sector, by district, 1990–1994

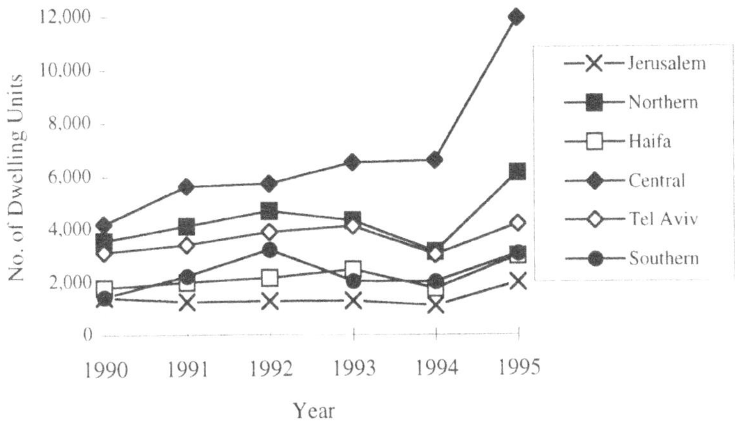

Fig. 25: Number of dwelling units built by the private sector, by district, 1990–1994

building at "emergency sites." Third, it permitted expedited building at other sites. Construction at "emergency sites" began in 1990, when it became clear that immigration to Israel from the Soviet Union would far exceed expectations. The Ministry of Construction and Housing concluded that a housing shortage might develop within half a year. As a result, the Ministry prepared an *emergency plan* based on building neighborhoods consisting of permanent one-story houses on the outskirts of existing towns. Some of the construction would be conventional (rigid building) and some would be wood-based. Relatively few apartments would be manufactured locally; most would be foreign imports. The main idea was to erect permanent housing quickly, on uniform plots of 250 square meters, that the residents could enlarge in accordance with plans provided to them when they took possession of the apartments. All the homes in the emergency plan were to be built or to arrive in Israel and to be habitable within a few months (Stern and Krakover 1994).

Most of the localities in which rapid building was planned were in the Northern and Southern districts; relatively few were in the Central District (see fig. 26). According to Stern and Krakover, the plans called for 57,000 dwelling units to be built in 63 locations. Of these, 31,000 were to be built in the Northern and Southern districts: 6,000 in ten locations in the Northern District and 25,000 in 27 locations in the Southern District. Only 10,000 dwelling units were actually built: 8,000 in 15 locations in the Southern District and 1,000 in four locations in the Northern District. The geographical distribution of expedited building sites shows a clear preference for the peripheral districts (for details, see Stern and Krakover 1994).

The government's spatial policy regarding immigrant absorption is marked by a certain internal contradiction. On the one hand, the direct-absorption policy and the policy of giving tax exemptions to landlords encourage immigrants—especially those of

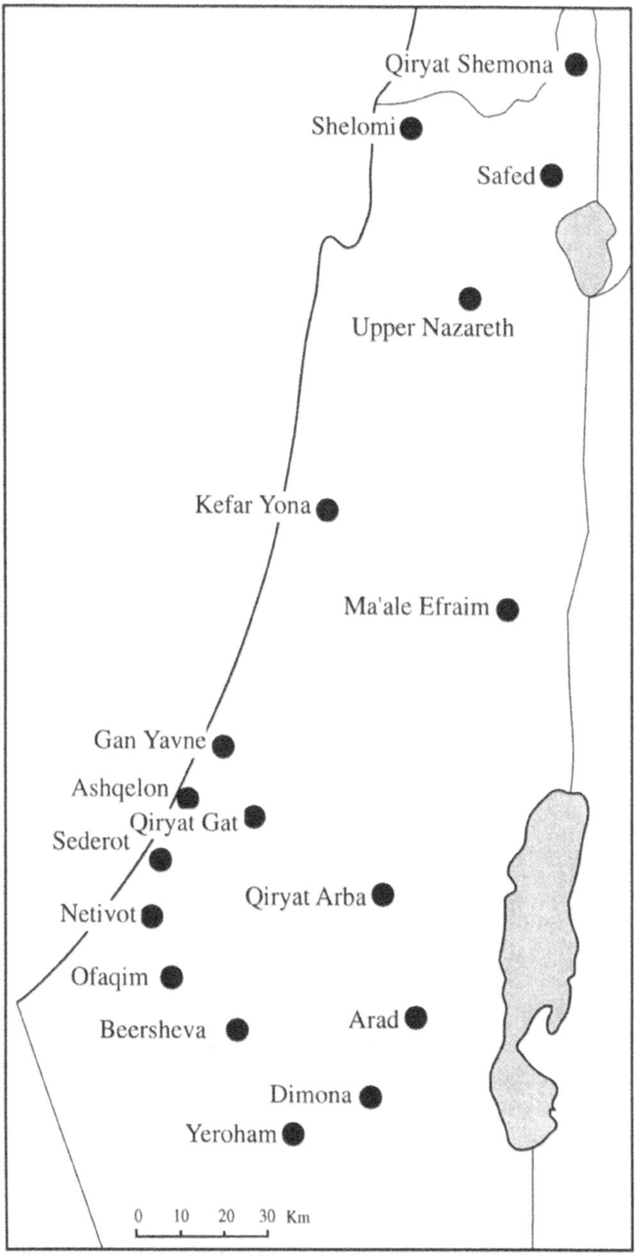

Fig. 26: Distribution of the sites slated for rapid building as of January 1993
(based on Stern and Krakover 1994)

working age and those with academic and scientific professions—to choose the central districts as their place of residence. In these districts home-purchase prices and rents are high, but the chances of finding work that suits their occupation and level of education are good. On the other hand, the policy of intervening in the spatial housing market by building huge numbers of apartments in the Northern and Southern districts and selling them at relatively low prices encourages immigrants to choose the peripheral districts, especially for housing reasons.

4.3 The Spatial Distribution of Immigrants' Residential Preferences: The Spatial Demand

The previous section described the spatial manifestation of housing supply: the public sector operates chiefly in the periphery and the private sector in the center. *What is the spatial manifestation of demand? Where do the immigrants choose to purchase homes?* It turns out that the immigrants' spatial behavior in purchasing homes reinforces the spatial behavior of the private market—a preference for central Israel. Further evidence of this is presented in table 5, which shows the geographical distribution of housing purchases by immigrants. The table shows clearly that the core districts (the Tel Aviv and Central districts) have the highest percentage of purchases; Haifa is in second place; Jerusalem and the south alternate in third place; and the north is last. In 1989–1994, for example, immigrants bought 91,000 apartments, 40 percent of them in the Tel Aviv and Central districts, 20 percent in the Haifa District, and 20 percent in the Northern and Southern districts. These findings show that a large percentage of new immigrants in the core districts see these areas as their permanent place of residence and therefore make the effort to buy an apartment. Many of the immigrants who live in the peripheral districts prefer to rent (rather than buy) apartments, because they are not certain that they will remain in those areas. Interestingly, the share of housing purchases by immigrants in central Israel was particularly high (over 50 percent) in 1989 and 1990; since

Table 5: Immigrants from the Former Soviet Union Who Purchased Homes with Housing Loans, by District of the Ministry of Construction and Housing

(percent)

Ministry district	Total (absolute numbers)	Total	1989	1990	1991	1992	1993	1994
Galilee (north)	8,969	9.8	1.7	5.8	7.7	10.4	11.8	8.9
Haifa	21,589	23.5	15.1	28.3	24.2	25.5	23.0	20.9
Center	35,362	38.6	54.9	50.9	39.5	35.6	34.7	43.4
Jerusalem	11,368	12.4	3.7	4.5	15.4	9.9	13.2	13.3
Negev (south)	12,953	14.1	24.0	9.9	12.2	16.2	15.6	12.1
Rural areas	1,222	1.3	0.5	0.3	0.7	2.1	1.4	1.2
Not classified	222	0.2	0.2	0.3	0.3	0.3	0.2	0.2
Total	91,685	100.0	100.0	100.0	100.0	100.0	100.0	100.0

then there has been a certain decline in the share of purchases in this district as a percentage of all home purchases in Israel, and the share in the Northern, Southern, and Jerusalem districts has increased. Most of the purchases were made in districts in which the public sector built little, and few purchases were made in the periphery, where the public sector builds a great deal.

What this means is that the immigrants who currently live in the core regions of Israel see their place of residence as permanent and therefore buy apartments whenever they can. Conversely, most of the immigrants in the periphery are fence-sitters. They generally refrain from buying homes, even though homes are relatively inexpensive. This indicates that there is a floating reserve of immigrants who still do not see the periphery as their permanent place of residence and think about moving from the periphery to central Israel. Although there has been a slight increase over time in the share who purchase homes in the peripheral districts as a percentage of all housing purchases in Israel, the basic trend still prevails.

Another manifestation of the immigrants' preference for the center over the periphery can be seen in table 6, which shows the geographical distribution of the immigrants as compared with Jewish nonimmigrants in 1990, 1991, 1992, 1993, 1994, and 1995.

Table 6 shows that approximately 65 percent of immigrants chose the core districts (Tel Aviv, Central, and Haifa) and only 25 percent chose the peripheral districts (Northern and Southern). In fact, *the new immigrants conformed to the spatial-distribution pattern of the nonimmigrant population*: they, like the nonimmigrants, prefer the core districts. For each year, the disparity between the share of nonimmigrant Jews and immigrants in each district (as a percentage of the total Israeli Jewish population and the total immigrant population, respectively) is fairly low (table 6). In 1990, for example, 27.7 percent of Jews in Israel lived in the Tel Aviv District, whereas 25.5 percent of immigrants lived there. In 1993, 10.3 percent of the Jewish population of Israel lived in the Northern District, while 10.0 of the total immigrant population lived there. The Central District was home to 24 percent of the total Jewish population and 21.3 percent of immigrants. There was a fairly large disparity between immigrants and nonimmigrants in the Haifa District in 1990, but it decreased over time.

In other words, although in the 1950s the geographical dispersion of the immigrants altered the geographical distribution of the Israeli population radically—as we saw in chapter 2—the immigration wave of the 1990s has hardly changed the pattern at all. The immigrants of the 1990s prefer the metropolitan areas, just as immigrants to Western countries do.

The number of immigrants in Judea-Samaria (the West Bank) and the Golan Heights is negligible. Of approximately 700,000 immigrants who have arrived in Israel since 1990, only 5,000 or so—less than one percent—live in these two regions.

A closer look at the data reveals that the immigrants who move to the Northern and Southern peripheral districts are essentially choosing towns that belong, from an administrative standpoint, to the Northern and Southern districts but geo-economically are part of the remote periphery of the Tel Aviv and Haifa metropolitan areas. These cities are situated on the "seam" between the national periphery and the metropolitan areas. In 1991, for example, about 65 percent of the 25,300 immigrants who settled in

Table 6: Immigrants and Nonimmigrants by District, 1990–1995

	1990		1991		1992		1993		1994		1995	
District	No.	Pct.	No.	Pct.	No.	Pct.	No.	Pct.	No.	Pct.	No.	Pct.
Jerusalem												
Nonimm.	427,100	10.8	444,300	10.7	455,100	10.7	464,000	10.7	473,200	11.0	482,100	10.6
Imm.	13,963	7.1	14,955	8.5	7,676	10.0	6,128	8.0	5,965	7.5	5,500	7.3
Northern												
Nonimm.	394,000	10.0	430,500	10.4	437,700	10.3	447,000	10.3	458,700	9.8	470,300	10.3
Imm.	20,839	10.5	28,041	15.9	7,879	10.2	7,777	10.2	9,201	11.6	9,400	12.4
Haifa												
Nonimm.	518,100	13.1	537,300	13.0	542,500	12.8	550,900	12.7	562,600	12.9	575,600	12.7
Imm.	42,966	21.7	29,454	16.7	12,179	15.8	12,566	16.4	13,743	17.6	13,300	17.6
Central												
Nonimm.	947,300	24.0	989,240	23.9	1,018,800	24.0	1,041,300	24.0	1,071,800	23.9	1,106,300	24.3
Imm.	45,857	23.2	36,697	20.9	15,861	20.6	15,007	19.6	14,826	18.7	14,000	18.6
Tel Aviv												
Nonimm.	1,078,700	27.4	1,112,300	26.7	1,119,300	26.4	1,115,800	25.8	1,115,400	27.7	1,116,300	24.5
Imm.	50,731	25.6	38,189	21.7	19,859	25.9	20,443	26.7	18,287	23.0	15,700	20.8
Southern												
Nonimm.	499,900	12.7	537,200	12.9	564,400	13.3	600,200	13.8	632,600	12.7	661,500	14.6
Imm.	22,189	11.2	27,097	15.4	12,626	16.4	13,795	18.0	16,416	20.7	16,600	22.0
Judea-Samaria												
Nonimm.	81,600	2.0	93,600	2.2	104,800	2.5	115,600	2.7	121,800	2.0	137,400	3.0
Imm.	1,462	0.9	1,594	0.9	861	1.1	862	1.1	946	1.2	1,000	1.3
Total												
Nonimm.	3,946,700	100.0	4,414,600	100.0	4,242,500	100.0	4,335,200	100.0	4,441,100	100.0	4,549,500	100.0
Imm.	198,007	100.0	176,027	100.0	79,941	100.0	76,578	100.0	79,384	100.0	75,400	100.0

Source: Central Bureau of Statistics, *Statistical Abstract of Israel*, 1991, 1992, 1993, 1994, 1995, 1996.
Note: Each cell in the table contains four numbers. In the Jerusalem District in 1990, for example, there lived 427,100 nonimmigrants, equal to 10.8 percent of the Israeli population. A total of 13,963 immigrants, or 7.1 percent of immigrants to Israel that year, made the Jerusalem District their initial place of settlement.

the Southern District lived in the three "seam" cities: Ashdod, Ashqelon, and Qiryat Gat. In the Northern District the number of immigrants in the "seam" cities of the Haifa region (Karmiel, Upper Nazareth, and Migdal Ha'emeq) is four times the immigrant population of the rest of the towns in the district.

Table 6 also shows a very small gap in the Southern District in 1990 and 1991 between immigrants and nonimmigrants; in 1992 and 1993 the gap was up to 3–4 percent in favor of the immigrants. It turns out that the massive government building that began in 1990 in the periphery—especially in the Southern District—produced notable spatial results: new immigrants who settled in the core districts moved to the periphery, especially to the Southern District.

However, the overall spatial result of the government's housing policy is, ultimately, a preference by immigrants for the core districts. Employment considerations and a sociocultural desire to live in or near the metropolitan areas outweighed housing considerations (low, subsidized prices in the periphery). Evidently, the immigrants are willing to live less comfortably—sometimes with several families in an apartment—in order to have access to jobs that suit their skills and to social and cultural centers (Lipshitz 1992a; Hasson 1993).

The immigrants are essentially faced with a spatial trade-off decision: they can choose a location where they have a good chance of finding a high-paying job that suits their professional skills and their level of education, but where housing prices are high; or a place where salaries are low and unemployment is high, but housing is inexpensive. In 1991 the average wage-earner in the core cities earned about NIS 2,500 per month, whereas the average wage-earner in the periphery earned only NIS 1,900 a month. In the same year, the unemployment rate was approximately 8 percent in the core cities and 13 percent in the peripheral districts.

Immigrants choose to live in the periphery only if they find jobs that suit their skills and level of education. Surveys conducted in development towns in the Israeli periphery indicate that most of the immigrants are satisfied with housing conditions there and are willing to remain provided that they find suitable jobs in these towns or within a one-hour commute. As long as such jobs are not available, they consider moving to central Israel (Lipshitz 1992a).

Thus, the geographical distribution of the immigrants of the 1990s did not alter the geographical distribution of the Israeli population. The immigrants of the 1990s replicated the preexisting spatial pattern, which was drawn in the late 1960s, when immigration slowed dramatically and virtually ceased. Figure 27, which depicts the

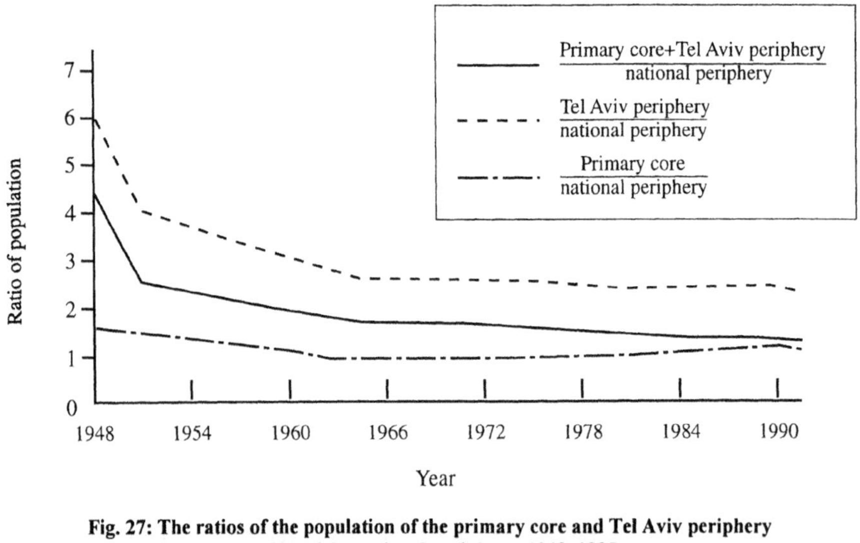

Fig. 27: The ratios of the population of the primary core and Tel Aviv periphery to that of the national periphery, 1948–1995

ratio of the population of the Tel Aviv core to that of the Israeli periphery (Northern and Southern districts), shows a substantial decrease in this ratio in the 1950s: from 6:1 to 3:1. This was a spatial result of the population-dispersion (immigrant-dispersion) policy in Israel on the national level. Since the late 1960s the ratio has remained almost constant, presumably as a result of the "equilibrium" between the higher rate of natural increase in the periphery and internal migration from the periphery to the core. Figure 27 shows that immigration from the former Soviet Union did not alter the core-periphery population ratio. Instead it reinforced the spatial distribution of the Israeli population, which was shaped in the late 1960s by internal migration in Israel (described in chapter 3).

Does the spatial pattern of the geographical distribution of immigrants' initial place of settlement remain constant over time? What is the spatial pattern of the immigrants' subsequent internal migration? Internal migration by immigrants from their initial place of settlement is a further manifestation of the spatial pattern of their housing demand. It expresses the immigrants' spatial preferences after they have become familiar with the economic-cultural spatial structure of their new country. Tables 7–8 and figures 28–33 show the spatial trends of internal migration among immigrants. It turns

Table 7: Internal Migration by Immigrants and Nonimmigrants: In-Migrants and Out-Migrants by District

District	1990		1992		1994		1995	
	Total	Thereof: immigrated since 1990	Total	Thereof: immigrated since 1990	Total	Thereof: immigrated since 1990	Total	Thereof: immigrated since 1990
In-Migrants								
Jerusalem	12,700	1,200	13,200	3,200	13,600	3,000	13,600	2,700
Northern	22,900	6,900	25,200	10,400	23,000	9,400	20,600	6,800
Haifa	24,300	6,200	27,000	10,600	28,700	11,700	27,200	9,600
Central	46,000	7,700	54,900	18,300	62,600	16,300	66,000	14,600
Tel Aviv	54,800	10,200	51,400	12,900	48,900	12,200	50,700	11,100
Southern	23,500	4,000	37,200	16,700	41,000	19,500	37,000	15,000
Judea-Samaria	9,800	600	14,000	2,200	13,500	2,500	13,100	2,400
National total	194,100	36,900	222,900	74,300	231,500	74,500	228,200	62,200
Out-Migrants								
Jerusalem	15,400	1,800	17,900	5,300	17,900	4,300	17,600	4,100
Northern	18,200	2,300	28,300	13,300	23,400	9,200	21,800	7,000
Haifa	28,900	9,000	31,400	12,500	29,600	11,200	27,800	9,100
Central	41,300	7,200	47,000	15,300	53,600	17,200	52,900	14,000
Tel Aviv	60,600	14,400	64,200	19,000	69,000	20,300	68,900	16,700
Southern	24,900	1,900	27,700	7,700	31,500	10,800	31,900	9,900
Judea-Samaria	4,800	300	6,400	1,100	6,600	1,400	7,300	1,400
National total	194,100	36,900	222,900	74,300	231,500	74,500	228,200	62,200

Source: Central Bureau of Statistics, *Statistical Abstract of Israel*, 1991, 1992, 1993, 1994, 1995, 1996.

Table 8: Net Internal Migration among the General Population and among Immigrants, by District, 1990–1994

	1990		1991		1992		1993		1994	
District	**A**	**B**	**A**	**B**	**A**	**B**	**A**	**B**	**A**	**B**
Jerusalem	–2,700	–600	–4,500	–1,400	–2,100	–4,700	–4,300	–2,000	–4,200	–1,400
Northern	4,700	4,600	5,600	4,800	–3,100	–2,800	–100	–500	–300	200
Haifa	–4,600	–2,800	–8,700	–6,000	–4,400	–2,000	–1,400	–100	–900	500
Central	4,700	500	–200	–2,300	7,800	3,000	2,500	–4,000	9,000	–900
Tel Aviv	–5,800	–4,200	–6,000	–2,600	–12,800	–6,200	–21,400	–10,000	–20,000	–8,100
Southern	–1,400	2,100	5,700	6,100	9,600	9,000	17,800	15,000	9,600	8,700
Judea-Samaria	—	—	8,000	1,400	7,600	1,100	6,900	1,600	6,900	1,000

A: Total net migration
B: Thereof: Immigrated since 1990
Source: Central Bureau of Statistics, *Statistical Abstract of Israel*, 1991, 1992, 1993, 1994, 1995.

out that internal migration alters the spatial distribution of the immigrants slightly in favor of the peripheral districts, especially the Southern District. The number of immigrants who leave the Tel Aviv, Jerusalem, and Haifa districts after initially settling there is greater than the number of immigrants who move to these districts, and in the Tel Aviv and Jerusalem districts the gap between the number of immigrant out-migrants and the number of immigrant in-migrants increases over time (tables 7–8 and figs. 28–33). In the Northern and Southern peripheral districts the picture is different. In the Southern District the number of immigrant in-migrants exceeds the number of immigrants who move to other districts from the Southern District, and this gap increases over time. In the Northern District the trend resembled that in the Southern District until 1991; then the trend was reversed, as more immigrants left the Northern District for other Israeli districts than moved there from elsewhere in Israel.

Table 7 depicts the migration of thousands of immigrants from the Tel Aviv, Central, and Haifa districts to the Northern and Southern peripheral districts.[1] In 1991, for example, net internal migration by immigrants was 4,800 in the Northern District, 6,100 in the Southern District, –4,900 in the Tel Aviv and Central districts, and –6,000 in the Haifa District. A similar pattern prevailed in 1990, too; internal migration by nonimmigrants increased population concentration, whereas that by immigrants tended to increase population dispersion.

The phenomenon ended in 1992, when net internal migration of immigrants remained positive in the Southern District but had become negative in the Northern District. Thus in 1992 internal migration added immigrants to the periphery but did not substantially change their spatial distribution, which shows a clear trend toward the metropolitan districts of Tel Aviv and Haifa.

Interviews with mayors in peripheral towns raise the possibility (which should

1. This was unexpected and even opposed to the general migration trend of immigrants in Israel, which is primarily from the periphery to the center (Lipshitz 1993a).

be checked empirically) that migrants from the center to the periphery are mainly "weak"—the elderly, the unemployed, and those in need of welfare services. These people prefer comfortable housing in the periphery to high rent in the center. The migrants also assume that welfare services for immigrants in the development towns of the Galilee and Negev are well developed. At the same time, a "strong population" leaves the periphery in the hope of obtaining jobs in the center that suit their technological skills (Lipshitz 1992a).

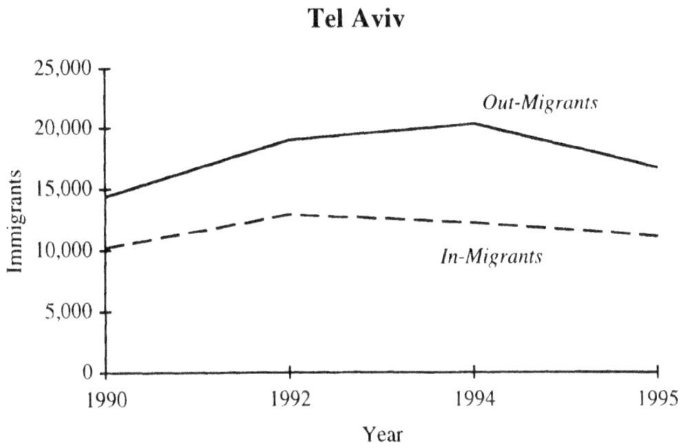

Fig. 28: Internal migration by immigrants in the Tel Aviv District, 1990–1995

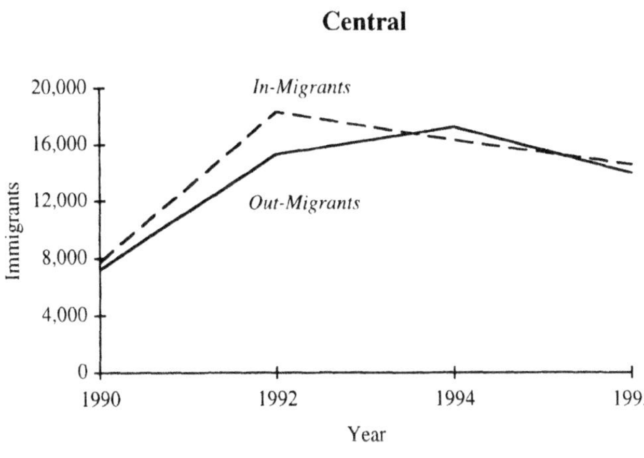

Fig. 29: Internal migration by immigrants in the Central District, 1990–1995

Thus, internal migration by immigrants, which modifies the spatial distribution of the immigrants slightly by increasing their share in the Northern and Southern districts at the expense of the core, simultaneously increases the percentage of "weak" immigrants in these regions somewhat.

In a study on the selection of place of residence by immigrants from the former Soviet Union, Hasson also states that the spatial distribution of immigrants leans slightly in the direction of the peripheral regions. He divides the periphery into two subregions: the Jerusalem and Haifa areas, and the Galilee and Negev (Hasson 1993). His surveys support the findings presented above regarding the gradual dispersion of

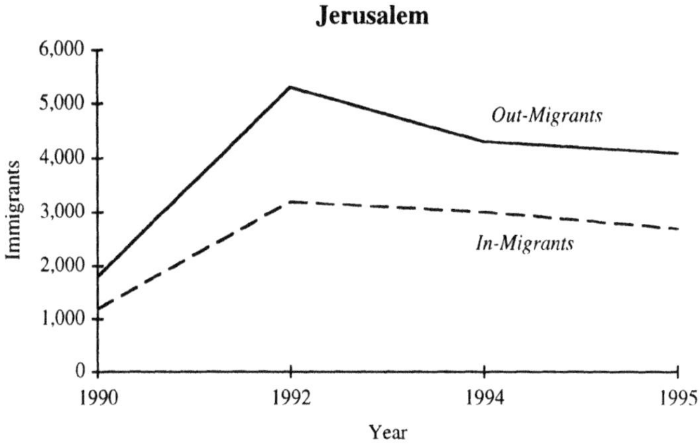

Fig. 30: Internal migration by immigrants in the Jerusalem District, 1990–1995

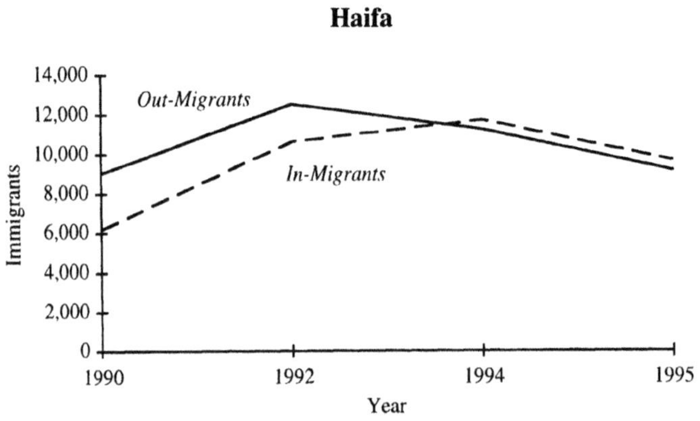

Fig. 31: Internal migration by immigrants in the Haifa District, 1990–1995

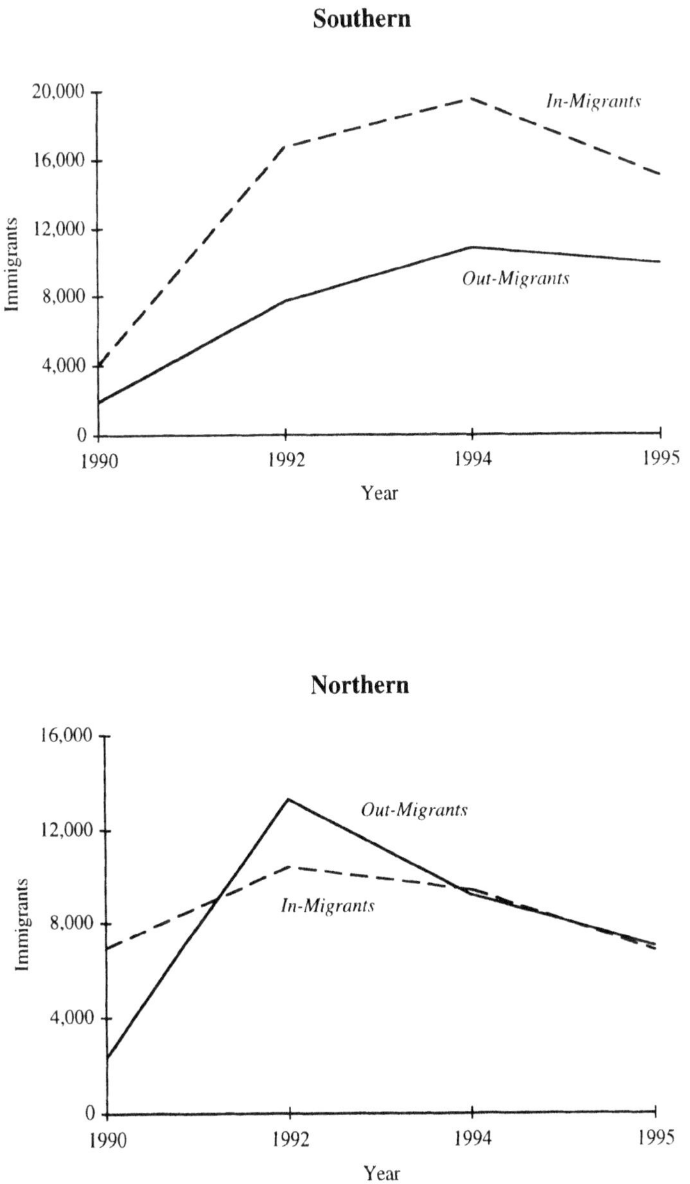

Fig. 32: Internal migration by immigrants in the Northern and Southern districts, 1990–1995

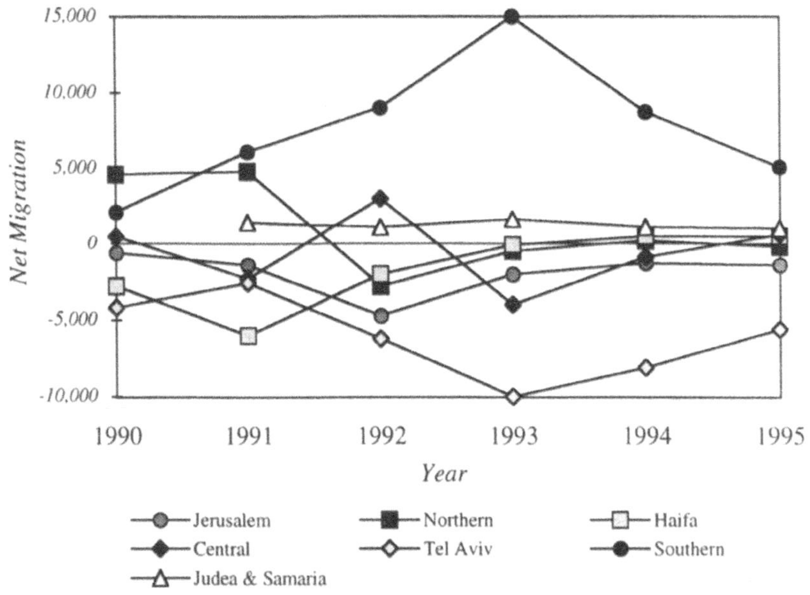

Fig. 33: Net internal migration of immigrants by district, 1990–1995

the immigrants to the periphery. Hasson states: "In creating a large supply of housing in the periphery, the government policy may have had a direct and indirect impact on the dispersion process" (Hasson 1993, p. 42). The direct impact was the availability of low-priced public housing for purchase. The indirect impact was downward pressure on rental prices due to the reserve of housing in the periphery.

What will long-term trends be like? Studies in Israel and abroad show that immigrants initially choose to live near relatives and friends in order to alleviate pressure and social and psychological crises. Over time, the need to be near relatives and friends lessens, and economic-employment considerations become more important (Lipshitz 1992a; Hasson 1993). The immigrants become willing to leave their supportive social network for jobs. The spatial distribution of the immigrants in the long term will thus be determined by the spatial distribution of suitable sources of employment.

Is there a spatial pattern describing the distribution of immigrants by republic of origin—especially the European republics versus the Asian republics? There do not seem to be any marked differences between the groups (see table 9). Immigrants from Russia, Belarus,[2] and Ukraine are represented in the Northern and Southern districts in slightly higher percentages than are immigrants from the Asian republics. For example,

2. To avoid confusion, the names used here for Soviet republics are the post-independence names, even when discussing the period before the breakup of the USSR.

Table 9: The Distribution of Immigrants by Republic of Origin and by District of Initial Settlement, 1990

	Russia, Ukraine, and Belarus	Moldova and the Baltic republics	Central Asian republics and Georgia	Total (absolute numbers)	Total (percent)
Jerusalem	77.4	4.5	18.1	11,877	100.0
Northern	76.8	11.9	11.3	21,027	100.0
Haifa	74.1	10.1	15.8	39,873	100.0
Tel Aviv	74.6	12.0	23.4	49,928	100.0
Central	73.7	6.3	22.0	44,804	100.0
Southern	71.0	11.4	17.6	16,689	100.0

Source: Ministry of Immigrant Absorption.

76 and 71 percent of immigrants who initially settled in the Northern and Southern districts, respectively, are from the European republics of Russia, Belarus, and Ukraine, whereas the average for the Soviet Union as a whole is 69 percent. Only 11 and 17 percent of immigrants from the Asian republics initially settled in the Northern and Southern districts, respectively, whereas the overall average for the Soviet Union is 20 percent.

Interestingly, a high percentage of immigrants from the Asian republics live in the Tel Aviv District. In general, on the national level the geographical distribution by area of origin in the 1990s is nothing like that of the 1950s, when the concentration of Asian immigrants in the national periphery was relatively high. Nevertheless, it is important to note that several development towns have very high concentrations of immigrants from the Asian republics.

4.4 The Geographical Distribution of the Immigrants on the Local Level

The previous section described and analyzed the geographical distribution of the immigrants by district, but it did not address the local level (the distribution of immigrants by locality). This section completes the analysis by considering the following questions: Is there a geographical pattern governing the distribution of immigrants by town? Do the immigrants tend to concentrate in certain localities within a certain district? What is the percentage of immigrants in each town in Israel, and do these percentages follow a geographical pattern? More specifically, what are the geographical patterns among the development towns in the Israeli periphery?

Table 10 and figure 34 show that immigration is unquestionably the dominant factor in the population growth of many development towns in the Israeli periphery. The Israeli population grew by an annual average of 4.2 percent in 1989–1993; the average annual rate of population growth in the development towns as a whole was greater. In some development towns, the average population growth rate was double or even triple the national average. For example, the population of Sederot increased from 9,700 to 14,600 within four years, for an average growth rate of 14.6 percent per year. Karmiel took in about 7,800 immigrants, boosting its population from 21,000 to 30,800, for an annual rate of growth of 11.7 percent. In some development towns, the contribution

Table 10: Population and Immigration, 1989–1993, by Town (Jewish population)

Town	Total population, 1989 (thousands)	Total number of immigrants	Total population, 1993 (thousands)	Average growth rate (percent)	Contribution of immigrants to population growth (pct.)	Immigrants as pct. of total 1989 population
Israel	4,559.6	529,478	5,327.6	4.2	68.9	11.6
Acre	37.4	6,436	45.1	5.1	83.6	17.2
Afula	25	3,802	33	8.0	47.5	15.2
Arad	13.8	4,255	18.3	8.2	94.6	30.8
Ariel	7	359	11.8	17.1	7.5	5.1
Ashdod	76.6	16,604	110.3	11.0	49.3	21.6
Ashqelon	56.8	9,550	73.3	7.3	57.9	16.8
Bat Yam	133.2	19,252	143.2	1.9	192.5	14.4
Beersheva	113.7	24,749	141.4	6.1	89.3	21.7
Beit She'an	13.1	452	14.8	3.2	26.6	3.4
Beit Shemesh	14.7	1,243	20.9	10.5	20.0	8.4
Bene Beraq	111.7	5,112	125	3.0	38.4	4.5
Dimona	24.8	2,707	30	5.2	52.1	10.9
Eilat	25.7	2,185	33.3	7.4	28.8	8.5
Givat Shemuel	10.0	603	10.7	1.7	86.1	6.0
Givatayim	45.6	2,783	47.2	0.9	173.9	6.1
Hadera	42.2	7,598	56.1	8.2	54.7	18.0
Haifa	223.6	58,741	246.5	2.6	256.5	26.2
Herzliyya	73.2	5,629	82.7	3.2	59.3	7.6
Hod Hasharon	24.8	1,499	27.7	2.9	51.7	6.0
Holon	148.3	17,675	162.8	2.4	121.9	11.9
Jerusalem	504.1	40,479	567.1	3.1	64.3	8.0
Karmiel	21	7,778	30.8	11.7	79.4	37.0
Kefar Sava	56.5	6,850	65.8	4.1	73.7	12.1
Lod	41.6	4,444	49.5	4.7	56.3	10.6
Ma'ale Adummim	12.7	830	16.9	8.3	19.8	6.5
Ma'alot-Tarshiha	8.8	2,719	12.8	11.4	68.0	30.8
Mevasseret Zion	11.8	1,487	14.4	5.5	57.2	12.6
Migdal Ha'emeq	15.3	3,274	20.1	7.8	68.2	21.4
Nahariyya	30.6	7,542	37.1	5.3	116.0	24.6
Nes Ziyyona	19.3	1,842	21.8	3.2	73.7	9.5
Nesher	10.5	379	16	13.1	6.9	3.6
Netanya	120.3	30,700	142.7	4.7	137.1	25.5
Netivot	10	938	13.6	9.0	26.1	9.4
Ofaqim	13.3	1,198	18.8	10.3	21.8	9.0
Or Aqiva	7.8	1,815	11.5	11.9	49.1	23.2
Or Yehuda	20.2	4,068	33.3	16.2	31.1	20.1
Pardes Hanna	16.3	1,842	19.4	4.8	59.4	11.3
Petah Tiqva	135.4	16,784	151.1	2.9	106.9	12.4
Qiryat Ata	36	4,877	41.8	4.0	84.1	13.5
Qiryat Bialik	32.8	3,738	35.3	1.9	149.5	11.4
Qiryat Gat	27.7	5,032	38.3	9.6	47.5	18.1
Qiryat Malakhi	14.6	1,650	18.6	6.8	41.2	11.3
Qiryat Motzkin	30.4	4,033	33.4	2.5	134.4	13.3
Qiryat Ono	22.2	1,911	23.8	1.8	119.4	8.6
Qiryat Shemona	15.3	2,210	19	6.0	59.7	14.4

continues ...

Town	Total population, 1989 (thousands)	Total number of immigrants	Total population, 1993 (thousands)	Average growth rate (percent)	Contribution of immigrants to population growth (pct.)	Immigrants as pct. of total 1989 population
Qiryat Tivon	11.8	1,024	12.5	1.5	146.3	8.6
Qiryat Yam	32.6	10,205	40.3	5.9	132.5	31.3
Ra'ananna	50.9	4,957	56.9	2.9	82.6	9.7
Ramat Gan	116.1	9,016	122.8	1.4	134.6	7.7
Ramat Hasharon	36.3	696	37.0	0.5	99.4	2.6
Ramle	45	5,927	55.5	5.8	56.4	13.1
Rehovot	73.8	14,952	83.2	3.2	159.1	20.2
Rishon Lezion	129.4	17,151	154.3	4.8	68.9	13.2
Rosh Ha'ayin	11.9	109	17.8	12.4	1.8	9.1
Safed	16.6	5,468	21.2	6.9	118.9	32.9
Sederot	9.7	1,816	14.6	12.6	37.1	18.7
Tel Aviv–Yafo	321.7	63,710	357.4	2.8	178.5	19.8
Tiberias	31.7	6,008	35.9	3.3	143.0	18.9
Tirat Hakarmel	14.6	1,212	17.2	4.5	46.6	8.3
Upper Nazareth	25.2	10,324	35.2	9.9	103.2	40.9
Yavne	21.3	1,518	25.6	5.0	35.3	7.1
Yehud	15.5	1,143	17.3	2.9	63.5	7.3

Source: Social science database, Hebrew University of Jerusalem (based on data from the Central Bureau of Statistics).

of immigration to population growth exceeded 100 percent; in other words, without the immigrants the population would have shrunk. In Upper Nazareth and Safed, for example, the contribution of immigration to population growth was 103.2 percent and 118.9 percent, respectively. Interestingly, in some development towns, such as Qiryat Gat, Beit Shemesh, and Sederot, immigration contributed only between 20 and 50 percent of their population growth. In these places it seems that immigration, which created the momentum for accelerated growth of the town's population and stimulated the local housing market, drew nonimmigrants from other localities, and these people also contributed to the town's growth. From the population growth trends in the development towns prior to the present wave of immigration (described in chapter 2) and the findings presented in this section, a clear picture emerges: the wave of immigration from the former Soviet Union extricated most of the development towns from a protracted deep freeze in population growth. Without these immigrants, the populations of many of the development towns would have remained stagnant over time because natural increase would have been offset by out-migration (Lipshitz 1990). Table 10 also shows that the contribution of immigration to population growth in the major cities was below the national average. For example, immigration contributed 2.8 percent to the growth of Tel Aviv, 2.6 percent to that of Haifa, and 3.1 percent to that of Jerusalem. A similar trend can be seen in towns in the periphery of the metropolitan areas, such as Ramat Gan, Givatayim, and Qiryat Motzkin. The growth rates of two cities in central Israel—Rishon Lezion and Netanya—exceeded the national average. Towns in central Israel whose demography and economic structure resemble those of the development towns have growth rates similar to those of the development towns. For example, Or Aqiva and Nesher have growth rates of almost three times the national average, and the growth rate of Or Yehuda is more than three times the national average (table 10). These towns had relatively large land reserves for construction because for years nonimmigrants had not been attracted to them. The immigrants chose to settle in these towns because

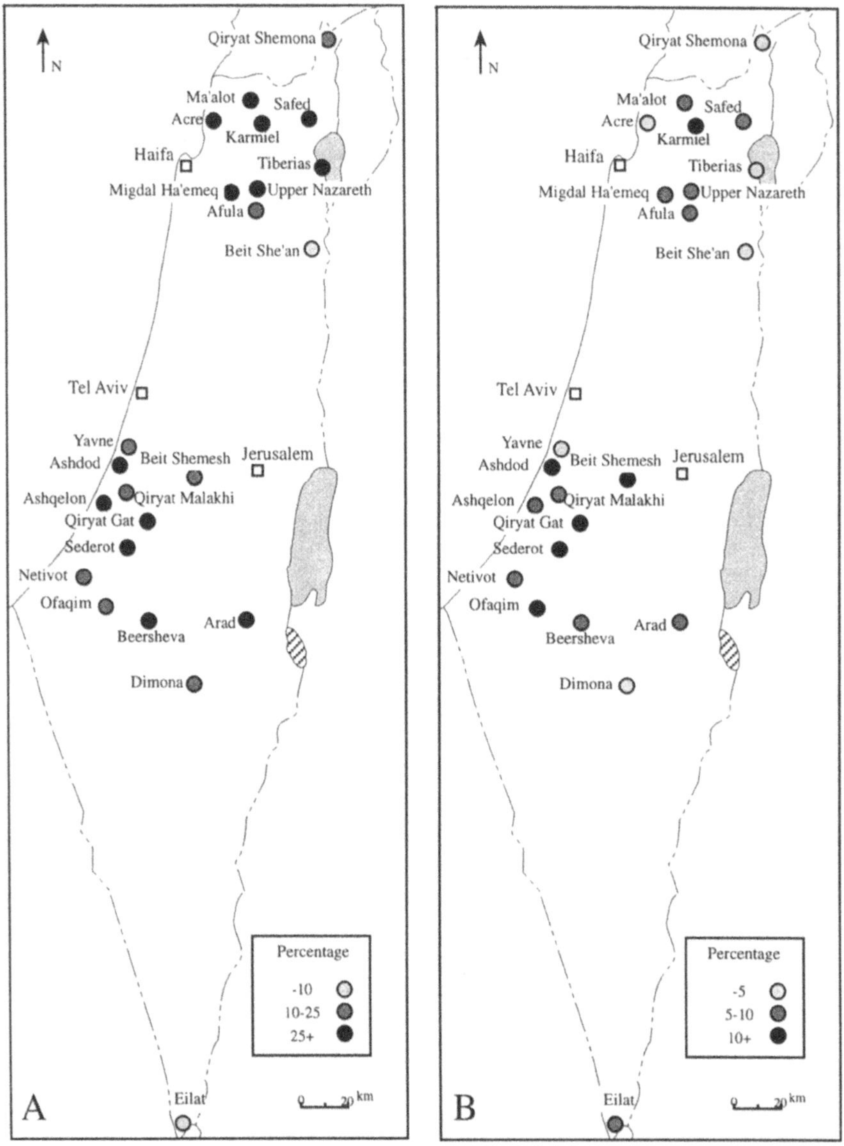

Fig. 34: The geographical distribution of the average population growth rate (A) and of immigrants as a percentage of the population (B) in peripheral development towns (1990–1994)

housing was relatively inexpensive and they were located near the job markets of the Tel Aviv and Haifa metropolitan areas.

Thus, although this wave of immigration did not change the distribution of the Jewish population significantly on the macro (national) level, and most of the immigrants settled in the core, on the micro level—the level of individual towns—the immigrants produced fairly substantial changes in the small towns (the peripheral development towns and towns in the central region, such as Or Aqiva and Or Yehuda, that were originally settled by immigrants in the 1950s). The immigrants produced relatively little change in the major cities of the central region.

In some development towns, immigrants account for a particularly high share of the population (see table 10 and fig. 34b). This is particularly notable in Upper Nazareth (40 percent), Karmiel (37 percent), Safed (32 percent), Arad (30 percent), Ma'alot (30 percent), Migdal Ha'emeq (21 percent), and Qiryat Gat (18 percent). These towns have one of two traits: either they are within commuting distance of the Tel Aviv or Haifa metropolitan area or they are on a medium to high socioeconomic level.

It should be noted that in some of the medium-sized and large cities in Israel, immigrants make up a significant percentage of the population. For example, the present immigrant population of Haifa is equal to 25 percent of the total 1989 population of the city; in Tel Aviv the present immigrant population is equal to 20 percent of the 1989 population. In Rehovot the number of immigrants who have arrived since 1990 is equal to about 20 percent of the 1989 population, and in Netanya the figure is 25 percent.

The fact that the immigrants are dispersed among the peripheral development towns does not necessary indicate that they regard these towns as their permanent place of residence. Two facts support this argument: First, Housing Ministry data (presented below) show that the bulk of home purchases by immigrants are in the central region and in the towns in the distant periphery of the Tel Aviv and Haifa metropolitan areas, such as Ashqelon, Ashdod, and Karmiel. There is also a relatively high rate of home purchases by immigrants in the development towns on the seam between the central region and the periphery, such as Migdal Ha'emeq, Upper Nazareth, and Qiryat Gat, as well as in the city of Beersheva. Second, a survey carried out among immigrants in five development towns shows that employment, not housing, is their prime concern. About 85 percent of those interviewed in Dimona, 72 percent in Sederot, 70 percent in Migdal Ha'emeq, and 80 percent in Qiryat Shemona responded that finding work is "the important" or "the most important" consideration in selecting a place of residence, and it will be the prime factor in any future move within the country. In other words, an absolute majority of persons interviewed believe that employment will be the main determinant of their future geographical location. About 80 percent of the immigrants in Dimona and 70 percent of the immigrants in Qiryat Shemona said that employment would be their main consideration in selecting their future place of residence; only 5 percent of the immigrants in Dimona and 10 percent of those in Qiryat Shemona cited housing as the main factor. In Qiryat Shemona and Dimona—as in Qiryat Gat—the presence of other Soviet immigrants and proximity to friends and family were important considerations in choosing their initial place of settlement. Approximately 54 percent of those interviewed moved to Qiryat Shemona and Dimona and 70 percent moved to Sederot and Migdal Ha'emeq because of information received from friends and relatives in

Israel; 25 percent of those interviewed in Qiryat Shemona and 31 percent in Dimona reported that they had been referred to those towns by absorption workers at the airport. In other words, the immigrants who moved to Qiryat Shemona, Sederot, Migdal Ha'emeq, and Dimona at the recommendation of family and friends would be prepared to leave this supportive social network—their relatives and friends—for suitable employment.

A sizable group of immigrants moved to Qiryat Shemona and Dimona on their own initiative (23 percent and 26 percent of immigrants in these towns, respectively) or at the recommendation of an Absorption Ministry worker (42 percent and 35 percent, respectively). Another 20 percent settled first in the central region and only later moved to these towns. These findings indicate free will to settle in Qiryat Shemona and Dimona, but according to the immigrants interviewed, the towns could become places of forced residence that they wish to leave unless the issue of suitable employment is resolved quickly.

It should be noted that other immigrants in Qiryat Shemona and Dimona never wanted to move to these towns at all; they would have preferred the central region or major Israeli cities. Furthermore, some immigrants (about 30 percent of those interviewed) report dissatisfaction with the town after a few months. When asked if they would like to remain in the town, 40 percent of the immigrants in Dimona and 35 percent of those in Qiryat Shemona responded, "I don't think so" or "definitely not." This group, which shows signs of instability, may influence the stable group.

Does the distribution of immigrants among development towns by republic of origin follow a spatial pattern? Table 11, which shows the initial place of settlement of Soviet immigrants in development towns in the Galilee and the Negev by republic of origin, indicates that some towns have particularly high percentages (above the national average) of immigrants from the European republics, while other towns have particularly high percentages of immigrants from the Asian republics. For example, Dimona and Qiryat Malakhi have high percentages of immigrants from the Asian republics (37.5 and 58 percent, respectively, compared to a national average of 20 percent), whereas Ofaqim and Arad have relatively high percentages of immigrants from the European republics. Interestingly, most of the northern development towns have higher than average percentages of immigrants from the European republics. For example, among immigrants who initially settled in Qiryat Shemona and Upper Nazareth, 75 and 79 percent, respectively, are from Russia, Belarus, and Ukraine; whereas the national average is 69 percent.

In general, it can be said that the distribution of immigrants by republic of origin in some development towns does not match the national average. For example, about 13 percent of Soviet immigrants in Israel as a whole are from the central Asian republics, as are 10 percent in the Southern District and 6 percent in the Northern District; however, 30 percent of the immigrants in Dimona and 18 percent of those in Qiryat Shemona are from these republics. About 48 percent of the immigrants who settled in Arad are from Belarus, whereas countrywide the average is about 24 percent. About half of the immigrants who settled in Qiryat Gat are from Ukraine, as compared with a national average of 31 percent. A study of immigrants in Sederot found that 34 percent are from the republics of the Caucasus and Azerbaijan, compared with a national average of 7 percent (Lithwick and Noam 1995). A survey carried out among immigrants in Qiryat

Gat found that 85 percent of the immigrants from Ukraine come from a single town—Donetsk. These trends call to mind immigration in other parts of the world. In countries that take in large numbers of immigrants, the newcomers tend to remain in close proximity to their social group, and therefore they live among others from the same country (Eisenstadt 1976).

In this respect, the immigrants from the former Soviet Union *have not followed the spatial patterns familiar from the pre-State period and the 1950s*. Then most European immigrants settled initially in the cities of the central region, whereas immigrants from Asia and North Africa settled in the peripheral development towns. Table 11 shows the concentration indices of the initial place of settlement of Soviet immigrants in development towns by republic of origin and compares these with the national average. The table shows clearly that in about half of the development towns the percentage of Soviet immigrants from European republics exceeds the national average. For example, 24 percent of immigrants in Israel in 1990 were from Russia, but 48 percent

Table 11: Concentration Index of Immigrants in Each Town by Main Republic of Origin, 1991
(national level=1)

Town	Russia	Ukraine	Belarus	Moldova	Baltic republics	Caucasus and Azerbaijan	Georgia	Central Asian republics
Acre	0.60	1.26	1.38	0.67	2.42	3.19	0.13	0.46
Afula	0.65	1.21	1.47	1.07	2.05	0.32	0.00	1.21
Arad	1.43	0.89	0.72	0.53	1.58	0.74	0.00	0.93
Ashdod	0.66	1.11	1.33	1.70	1.15	0.70	0.71	1.00
Ashqelon	1.12	1.15	1.31	0.96	0.79	0.73	0.61	0.47
Beersheva	0.92	0.97	0.70	1.61	1.54	1.81	0.72	0.79
Beit She'an	0.95	1.19	0.47	0.00	0.00	0.00	3.45	2.22
Beit Shemesh	2.06	0.40	0.22	0.45	0.00	0.48	0.00	1.21
Dimona	0.70	0.61	0.89	0.34	0.00	1.16	0.00	3.14
Eilat	0.67	1.51	1.87	0.00	2.06	0.00	4.12	0.81
Hatzor Hagelilit	0.00	0.00	0.00	0.00	0.00	0.00	0.00	0.00
Karmiel	0.98	1.26	1.23	1.08	0.71	0.85	0.52	0.54
Ma'alot	1.04	1.12	2.65	0.41	1.02	0.00	0.00	0.26
Migdal Ha'emeq	0.57	1.16	0.84	2.65	0.77	2.10	0.00	0.49
Mitzpe Ramon	0.77	0.60	0.90	0.00	0.00	1.13	0.00	3.22
Netivot	1.14	0.80	3.79	0.23	0.00	0.00	0.00	0.00
Ofaqim	1.10	0.30	4.55	0.29	0.00	1.05	0.00	0.00
Qiryat Gat	0.95	1.08	1.85	1.77	0.00	0.23	0.57	0.31
Qiryat Malakhi	1.31	0.79	0.94	0.35	0.00	0.95	2.38	1.30
Qiryat Shemona	0.89	1.44	0.81	0.64	0.00	0.00	0.00	1.43
Safed	1.58	0.86	0.10	0.40	2.40	0.43	0.00	1.22
Sederot	0.30	0.95	0.84	4.10	0.00	1.54	0.00	0.88
Shelomi	0.42	0.50	1.58	0.48	2.17	2.61	0.00	2.48
Tiberias	0.95	1.12	1.67	0.50	0.53	0.90	0.00	0.93
Upper Nazareth	0.71	0.95	2.53	1.45	0.65	0.52	0.00	0.60
Yeroham	0.65	0.48	0.45	0.83	0.00	1.50	7.50	2.86
National total	**15,882**	**13,005**	**5,731**	**4,533**	**1,245**	**2,684**	**403**	**7,245**

of immigrants in Arad, 37 percent of those in Beit Shemesh, and 45 percent of those in Safed were from this republic. Despite these figures, the immigrants from the European republics are not concentrated in specific development towns nearly as much as the immigrants from the Asian republics are. In 1991, for example, the percentage of immigrants from the central Asian republics was 3 times the national average in Dimona, 2.9 times the national average in Yeroham, and 2.5 times the national average in Shelomi; in 1990 the percentage from the central Asian republics in Qiryat Malakhi was 4 times the national average. In 1990 the percentage of Georgian immigrants in Netivot was 3.9 times the national average, that in Eilat was 4 times the national average, and that in Yeroham was 7.5 times the national average. In 1990, the concentration of immigrants from the Caucasus and Azerbaijan was particularly high in Yeroham and Sederot: over 3 times the national average.

In conclusion, the government's direct and indirect housing policy and the immigrants' behavior in the free housing market had positive effects. A study of immigrants who had arrived in July 1990 and had been in Israel for at least five years indicates that there has been a significant improvement in the percentage living in permanent housing (Damian and Rosenbaum-Tamari 1996). Seventy percent reported that they live in homes owned by the family, and 9 percent live in public or government-owned housing. In other words, approximately 80 percent of the immigrants live in permanent housing. There has also been a substantial decline in housing density over time: from 1.4 persons per room one to two years after immigration to 1.1 persons per room after five years.

The same study shows that the level of satisfaction of July 1990 immigrants with their town and neighborhood is much higher than in each of the previous years. Furthermore, the immigrants become more and more satisfied with their housing the longer they remain in Israel: 60 percent were satisfied with their housing after two years in Israel, whereas 80 percent were satisfied after five years (Damian and Rosenbaum-Tamari 1996).

The geographical distribution of immigrants from the former Soviet Union closely resembles the distribution of nonimmigrants, with a slight preference for the Southern District. This distribution has done very little to change the geographical distribution of the Israeli population.

The macro-level analysis indicates that when immigrants' selection of a place of residence is influenced chiefly by market forces (direct absorption) rather than government intervention (spatial policy), they decide the same way immigrants to Western Europe and North America do. Immigrants to Israel—like immigrants to other countries—favor the core regions, where job prospects are good. Their preference for the Tel Aviv and Haifa metropolitan areas recalls to a certain extent the preference of immigrants for "global cities" (even though Tel Aviv and Haifa do not exactly qualify as global cities).

The spatial preferences of the housing demand of immigrants correspond to the spatial supply of apartments on the private market. As the findings showed, the private market clearly preferred residential building in the core regions. Conversely, the spatial supply of the public market was more prominent in the peripheral regions, where substantial reserves of public land were available.

Government residential building in the peripheral districts, and especially in the Southern District, at fairly low prices and with substantial subsidies, prompted immigrants to move to the peripheral districts. Many immigrants who initially settled in the Tel Aviv, Central, and Haifa districts moved to the peripheral districts, where they joined immigrants who had chosen these districts as their initial place of settlement. It is important to keep in mind, however, that the rate of housing purchases by immigrants is high in the Tel Aviv, Central, and Haifa districts and low in the peripheral districts. In other words, while immigrants show signs of residential permanence in the core districts, in the peripheral districts a temporary feeling is dominant.

A closer look at the data (not presented in this book) shows that the preferred regions and towns in the peripheral districts are those located on the seam between the periphery of the core and the national periphery (the peripheral districts), such as Ashdod, Ashqelon, Qiryat Gat, Migdal Ha'emeq, and Karmiel. With improvements to the transportation infrastructure, some of these towns have become part of the distant periphery of the Tel Aviv and Haifa metropolitan areas. Immigrants and nonimmigrants alike in these towns enjoy two advantages of location: relatively low housing prices and reasonable access to metropolitan job centers. The remote towns in the national periphery do not attract immigrants who previously lived in the center, and those who do move to them tend to be socioeconomically weak. This trend harms the economic growth potential and the quality of welfare services.

CHAPTER 5

THE ECONOMIC ABSORPTION OF IMMIGRANTS FROM THE FORMER SOVIET UNION IN THE 1990s: THE NATIONAL AND REGIONAL LEVELS

This chapter examines the economic aspects of the absorption of immigrants from the former Soviet Union in Israel on two geographical levels: the national level and the regional level. On the regional level, emphasis will be placed on the Israeli peripheral regions, where the hiring of immigrants (or the immigrants' desire to be hired) has been strongly felt.

5.1 The Absorption of Immigrants from the Former Soviet Union in the Israeli Labor Market: The National Level

In the first three years of the 1990s, the Israeli economy achieved high rates of growth unknown since the early 1970s. This rapid growth was spurred by a large wave of immigration that began in late 1989, and was based on economic reforms introduced in 1985 (Bank of Israel 1994). The effect of immigration on the economy changed over time. For example, factors related to initial absorption, such as absorption baskets and the immediate need for housing, became less important, and factors related to adaptation of economic behavior patterns to those of the general Israeli population in terms of consumption of durable products, employment, and housing became more important. The Bank of Israel report states that "at the end of the fourth year of the large wave of immigration, the extent of the economic absorption of the immigrants in Israel can be summed up as a partial success (Bank of Israel 1994, p. 3; see also Beenstock and Ben-Menachem 1995). The rate of immigrant participation in the labor market is high, and the unemployment rate among the immigrants dropped from 40 to 20 percent. The general unemployment rate in Israel was 8.2 percent in the first quarter of 1989, but had risen to 9.2 percent by the last quarter of the year. In 1990 unemployment was up to 9.7 percent, and it rose to 11 percent in 1991 and 11.3 percent in 1992, before dropping to 10.0 percent in 1993 and to 7.8 percent in 1994. During this period, too, real wages of wage-earners in the business sector dropped by 6 percent altogether. The decline in hourly wages was greater. The drop in real wages primarily reflects the effect of the increase in unemployment. Real wages in the public sector continued to rise during this period, but at a moderate pace. Due to the age structure of immigrants from the former Soviet Union, a large number of immigrants joined the labor force within a short period of time. The main dilemma that confronted and still confronts policymakers in this field is how to create jobs, especially jobs that suit the skills and occupations of the immigrants.

The present wave of immigration is marked by high-caliber human capital, as manifested in a much higher proportion of university graduates with technical and medical professions than in the Israeli labor force. "Their integration in Israel will make the country's labor force the most educated in the world" (Ofer, Plug, and Kasir 1991, p. 135). The problem, however, is that the structure of the Israeli economy is not prepared in the short term to take in such highly developed human capital. Due to the disparity between the supply of and demand for a labor force with a relatively high percentage of university graduates, the researchers foresee a high rate of change and downward adjustment of the immigrants' occupations (Ofer, Plug, and Kasir 1991).

The country thus had a rare opportunity to effect a structural change in its economy so as to take advantage of its relative strength: educated human capital. A large proportion of the immigrants are university-educated and have backgrounds in technical fields; these people had (and have) the potential to turn economic development in the direction of sophisticated, high-tech industry, including chemistry, metals, electricity, and electronics. It was not inconceivable that we would see modern changes in the manufacture of traditional products, too. Therefore the absorption problem is not only a matter of expanding sources of employment, but of expanding them in such a way as to take the immigrants' skills into account. In other words, the maximum development effort should be put into expanding industries and services that require human, scientific, engineering, technical, and (if possible) medical capital.

It turns out that the economy was not able to take advantage of the educated and skilled newcomers. Indeed, there were substantial difficulties in absorbing the immigrants in accordance with their academic and technological qualifications. The economic growth that occurred was in a direction that is not desirable in the long run (Plug and Kasir 1994). In the amount of time that it took earlier immigrants to find work in their original fields, only a small portion of the recent immigrants had done so, especially in fields that call for high-caliber human capital. Quite likely, this situation is one of the reasons for the decline in the rate of immigration since 1990 of people with scientific professions and higher education. Among 1990 immigrants, 39 percent worked in scientific professions and 16 percent worked in industry; in 1993, approximately 27 percent worked in scientific professions and 26 percent were industrial workers. Among 1990 immigrants, 57 percent had at least 13 years of education; among 1993 immigrants the figure was down to 43 percent (CBS 1994).

Table 12 illustrates the high quality of human capital that reached Israel during the wave of immigration from the former Soviet Union. It shows that 50–60 percent of the immigrants aged 15 and over have had at least 13 years of education, whereas the Israeli figure is 30 percent; 35–40 percent of the immigrants in the labor force have academic or scientific occupations, compared to 12 percent of Israelis overall; 30–35 percent have technical occupations, compared to 14 percent of Israelis. The percentage of immigrants from the former Soviet Union with higher education and scientific or academic professions dropped from 39 percent in 1990 to 27 percent in 1993 (table 12).

Even if we assume that some of the reports regarding the immigrants' education and occupations are untrue, there is no doubt that *the immigration from the Soviet Union is an import of high-caliber human capital.* Under normal conditions a country would have to invest extensive resources to attain such a level of human capital. In an era in which high-tech industry and computer and information services are important elements in economic growth, this immigration wave brought great potential for the

Table 12: Distribution of Occupations and Education of Immigrants, 1990–1993*

Occupation*	1990 immigrants	1991 immigrants	1992 immigrants	1993 immigrants	1990–1993 immigrants
Scientific professions	39.0	35.5	32.6	27.1	35.3
Thereof: Doctors	6.2	4.3	3.8	3.4	4.8
Engineers	25.4	22.3	19.9	15.9	22.3
Other professional, technical, and related workers	34.3	32.4	32.7	33.3	33.3
Industrial workers (skilled and unskilled)	15.7	19.1	23.2	26.4	19.3
Education of those aged 15 and over					
9–12 years	31.9	35.0	39.0	43.1	35.4
13 or more	57.3	52.7	48.8	43.1	52.7

*Occupation as declared by the immigrant.
Source: Bank of Israel report, 1993 (based on data from the Central Bureau of Statistics).
Note: The percentages do not add up to 100 because not all categories are included in the table.

growth of the Israeli economy. So far, however, it is the scientific and technological professionals who have had the hardest time finding appropriate work. Figure 35 shows clearly the *gap between the distribution of the immigrants from the former Soviet Union by their occupation in their country of origin and in Israel.* As it turns out, many of the immigrants found work in the services and as production workers in traditional industry; only a few found jobs that suit their academic and scientific skills. About 33 percent of the immigrant men have academic and scientific professions, and in 1991 only 7 percent of them worked in this occupation; among women the gap is 35 percent. The services, as figure 35 shows, employed 3 percent of the men and 5 percent of the women in the Soviet Union, but 15 percent and 45 percent, respectively, in Israel.

With respect to the disparity between supply of appropriate jobs and demand for them, Herkovich and Meridor note that the chief economic boon of the immigrants' high-quality human capital is in the long term, while in the short term problems can be expected in various spheres, especially in employment and housing (Herkovich and Meridor 1991). On the same subject, Ben-Bassat argues that employment is the main issue "over which the immigration will rise or fall" (Ben-Bassat 1991, p. 267). In his opinion, the immigrants have proven that they are willing to compromise—and even to make major compromises—on housing. They are not, however, willing to compromise with respect to the lack of jobs in general, and often not with respect to the lack of jobs that suit their occupations and level of education. From the studies surveyed it is evident that the main dilemma on the national level was, and still is, how to create jobs that suit the immigrants' academic, scientific, and technical occupations. Partial failure to take advantage of the human capital of the immigrants who had arrived by the end of 1993 produced a loss of $2 billion to the national economy (Eckstein 1994).

It should be noted that the percentage of immigrant academics and scientists who work in these fields in Israel increases with time in Israel (see fig. 35). As figure 35 shows, 30 percent of adult male immigrants who arrived in October-December 1990 worked in academic and scientific professions in the Soviet Union, whereas only 7 percent were doing so after 1.5 years in Israel (January-April 1992). After 2.5 years in

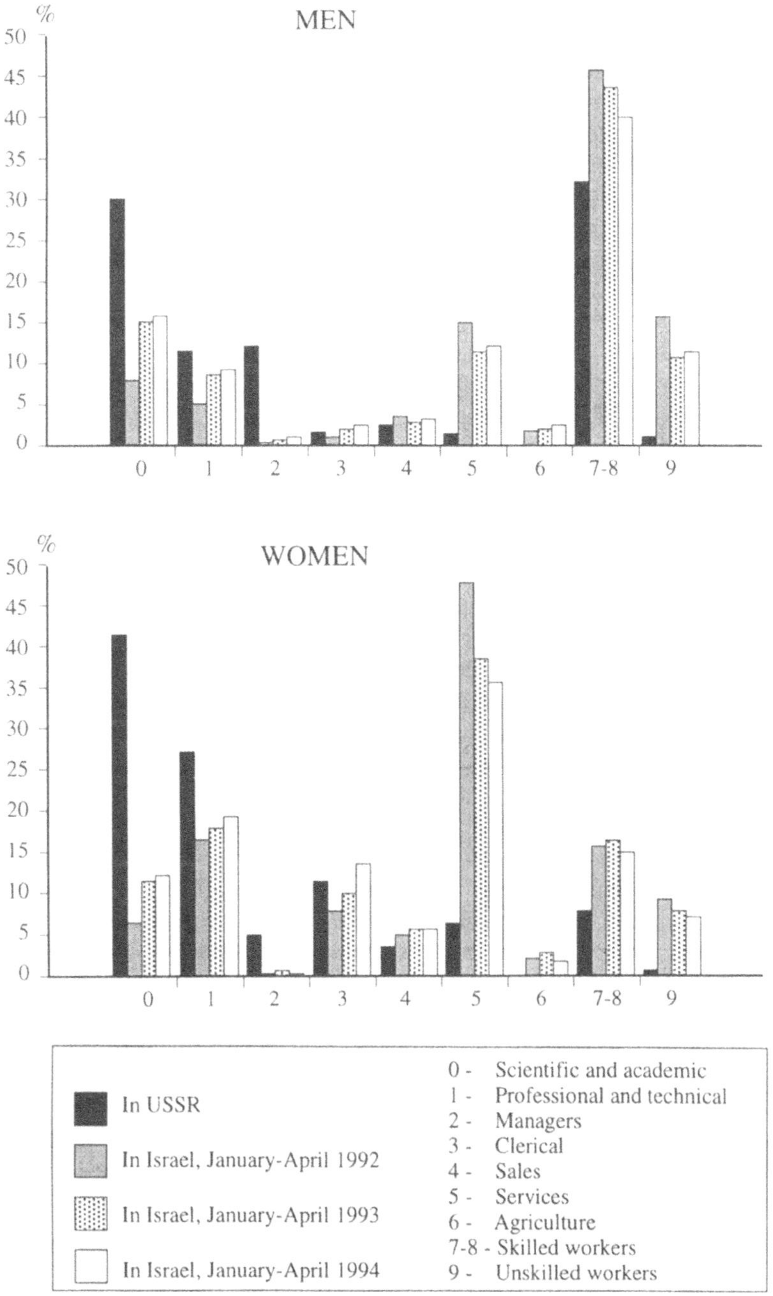

Fig. 35: Occupational distribution in the former Soviet Union and in Israel, 1992, 1993, and 1994 (source: Central Bureau of Statistics, "Employment of Immigrants from USSR Who Arrived in Israel in October-December 1990: A Follow-Up Survey, Three Years after Immigration," *Supplement to the Monthly Bulletin of Statistics*, no. 6, 1995)

Israel (January-April 1993), 15 percent of the immigrants worked in academic and scientific professions, and by January-April 1994 (3 to 3.5 years after their arrival) the figure was up to 16 percent. Concurrently, the share employed in the services dropped. A similar trend was found among women. In other words, the longer immigrants have been in Israel, the greater is the percentage employed in academic and scientific professions and the fewer are employed in the services (see fig. 35). However, the disparity between the percentage of male and female immigrants with scientific and academic professions who found work in these fields is highly significant.

The time factor is significant in finding work in Israel. Figure 36 shows clearly that the immigrant unemployment rate decreases substantially over time. For example, about 30 percent of women aged 35–44 who moved to Israel in October-December 1990 were unemployed 1–1.5 years later (January-April 1992), whereas only 18 percent were unemployed after 2–2.5 years (January-April 1993) and only 10 percent after 3–3.5 years (January-April 1994). Figure 36 also shows that despite the decline in the unemployment rate of each group with time in Israel, the disparity between age groups remains. For example, women immigrants aged 35–44 have a higher unemployment rate than women immigrants aged 25–34. As figure 36 shows, there is a substantial gap between the unemployment rates of men and women, but this disparity narrows over time, especially due to the dramatic decrease in the unemployment rate among women the longer they have been in the country.

The improvement in the unemployment rate, especially the increase in the percentage of immigrant scientists and academics employed in their fields, also boosts the immigrants' income level. *Immigrants' incomes increase the longer they are in Israel:* average net monthly income from all sources (including government financial support and support by parents who receive old-age pensions and live with their children) was NIS 1,640 among immigrants who had been in Israel for half a year to a year, and NIS 2,710 among immigrants who had been in Israel for two years or more. The average net income of immigrant families headed by wage-earners was NIS 2,770 a month; it rose to NIS 3,030 a month among those who had been in Israel for two years or more.

Studies that have looked specifically at the success of the absorption of immigrant engineers (Naveh, Noam, and Benita 1993) and immigrant doctors (Nirel and Naveh 1996) in the Israeli job market show that, in general, the doctors have done better than the engineers. In a study of 6,754 immigrant doctors who came from the former Soviet Union between 1990 and 1992, Nirel and Naveh report that 72 percent of those licensed to work as doctors in Israel are doing so, a minority of them in temporary jobs. Furthermore, most of the doctors work for public employers (Health Ministry, health funds, and public hospitals), and can therefore expect job stability and fair pay for their work.

Among engineers the picture is different. A study based on interviews with 1,500 immigrant engineers from 1989 to 1994 (Naveh 1996) found that only 21 percent of them work as engineers; the rate is higher among men (27 percent) than women (14 percent). The percentage of immigrant engineers working as engineers increases with time in Israel, but the improvement is not dramatic: only 32 percent of immigrant engineers who have been in Israel for five years work as engineers. The study also found that the vast majority of immigrant engineers who do not work as engineers do not have jobs that suit their level of education.

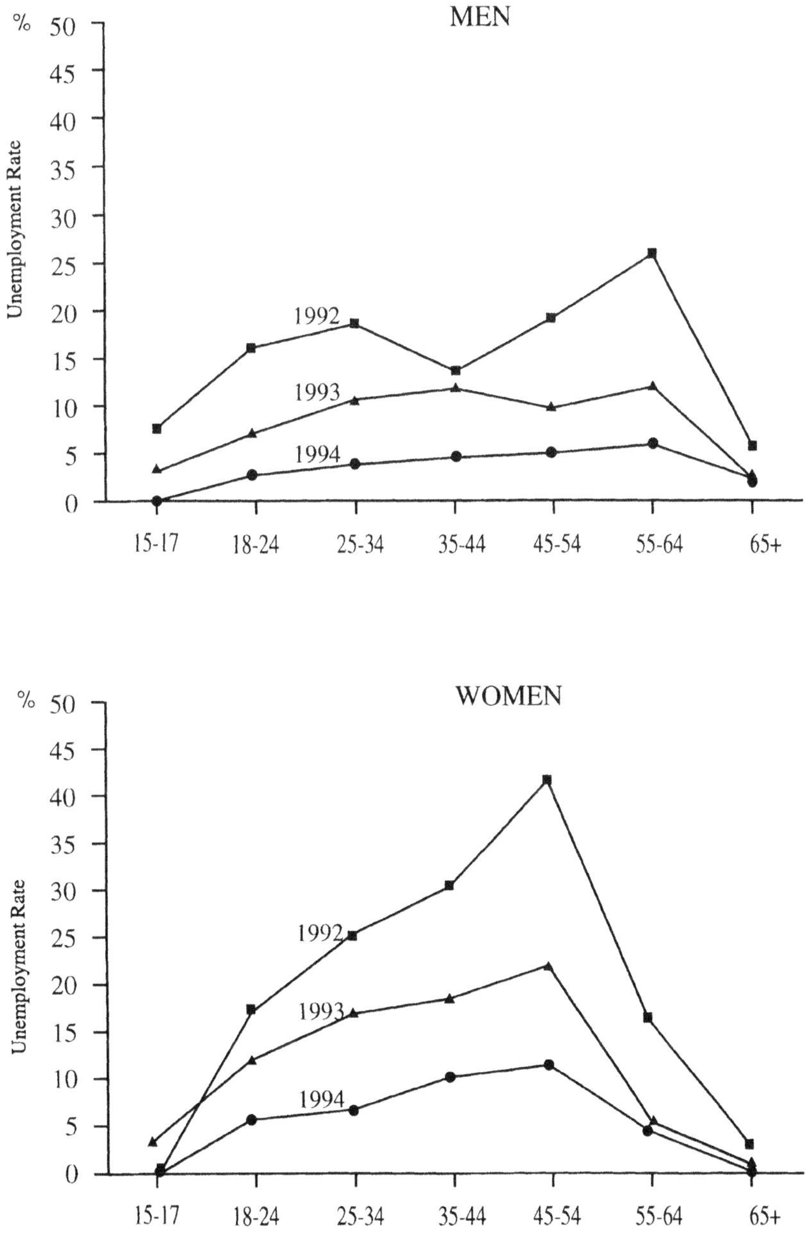

Fig. 36: Unemployment in January-April 1992, 1993, and 1994, by age (immigrants from the USSR who arrived in October-December 1990) (source: Central Bureau of Statistics, "Employment of Immigrants from USSR Who Arrived in Israel in October-December 1990: A Follow-Up Survey, Three Years after Immigration," *Supplement to the Monthly Bulletin of Statistics*, no. 6, 1995)

The dramatic drop in immigrant unemployment rates substantially reduced the gap between the unemployment rates of immigrants and nonimmigrants (fig. 37). In 1990, 35 percent of immigrants were unemployed, as against 8 percent of nonimmigrants; the disparity was thus 30 percent. In 1993 the disparity was down to about 7 percent, as immigrant unemployment had fallen to 13 percent and that of nonimmigrants to 6 percent (see fig. 37). The disparity between the unemployment rates of immigrant and nonimmigrant women in 1994 was 8 percent (see fig. 38) and that between immigrant and nonimmigrant men was 5 percent (see fig. 39). Despite the dramatic decline in unemployment rates among immigrant women—from 52 percent in 1991 to 18 percent in 1994—unemployment among immigrant women is still consistently higher than among immigrant men (figs. 38 and 39).

It is interesting to note the decline in unemployment rates among immigrant university graduates. The figure rose from 32 percent in 1990 to 34 percent in 1991. Then a sharp decline in the unemployment rates of immigrant university graduates began (fig. 40). In 1991 the gap between unemployment rates among immigrant and nonimmigrant university graduates was 30 percent; in 1993 it was only 10 percent. Presumably, some of the decline among the immigrant university graduates is because they accepted jobs in non-academic fields. It should be noted that the disparity in unemployment rates between immigrants with no university degree and nonimmigrants with no university degree has decreased more than that between university graduates; in 1993 it was 5 percent (see fig. 41). The successful economic absorption of the immigrants is manifested in part in a decrease in the income disparities between immigrants and nonimmigrants (fig. 42).

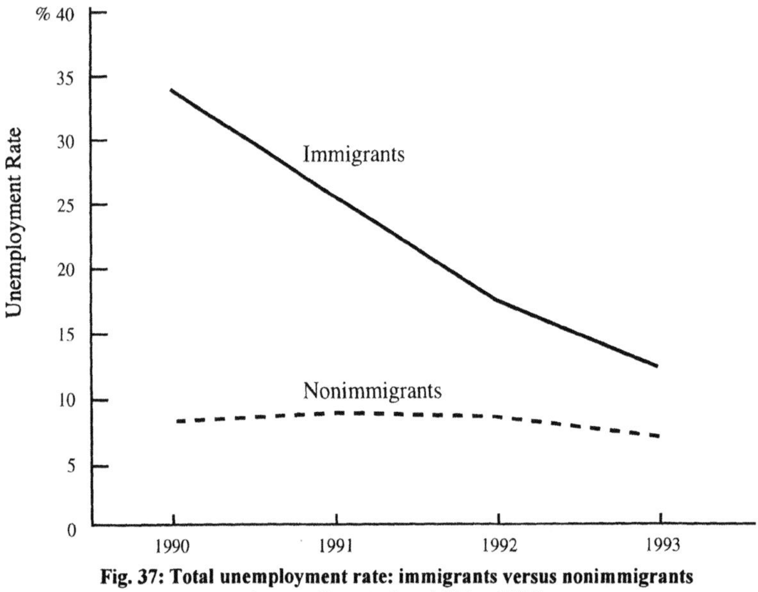

Fig. 37: Total unemployment rate: immigrants versus nonimmigrants
(source: Beenstock and Fisher 1996)

The findings presented above indicate that the main dilemma *on the national level* was and remains how to create sources of income that suit the academic, scientific, and technical occupations of the immigrants. What are the geo-economic trends regarding

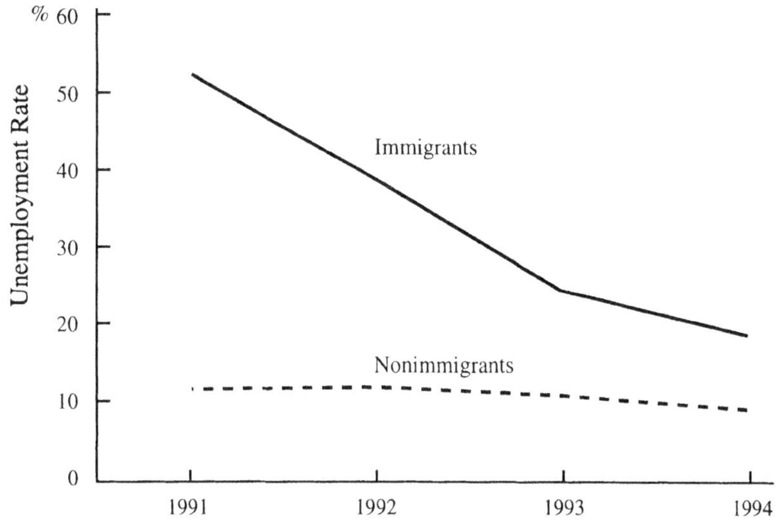

Fig. 38: Unemployment among women: immigrants versus nonimmigrants
(source: Beenstock and Fisher 1996)

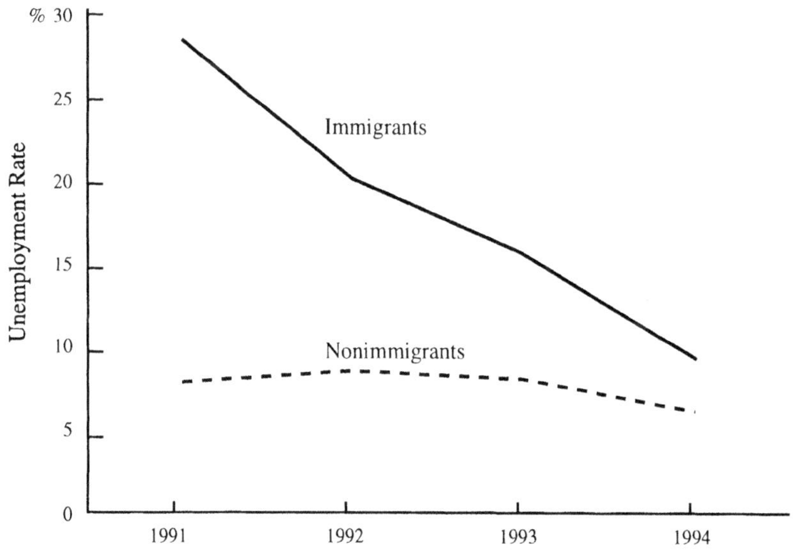

Fig. 39: Unemployment among men: immigrants versus nonimmigrants
(source: Beenstock and Fisher 1996)

the absorption of immigrants from the former Soviet Union in Israel? This is discussed in the next section.

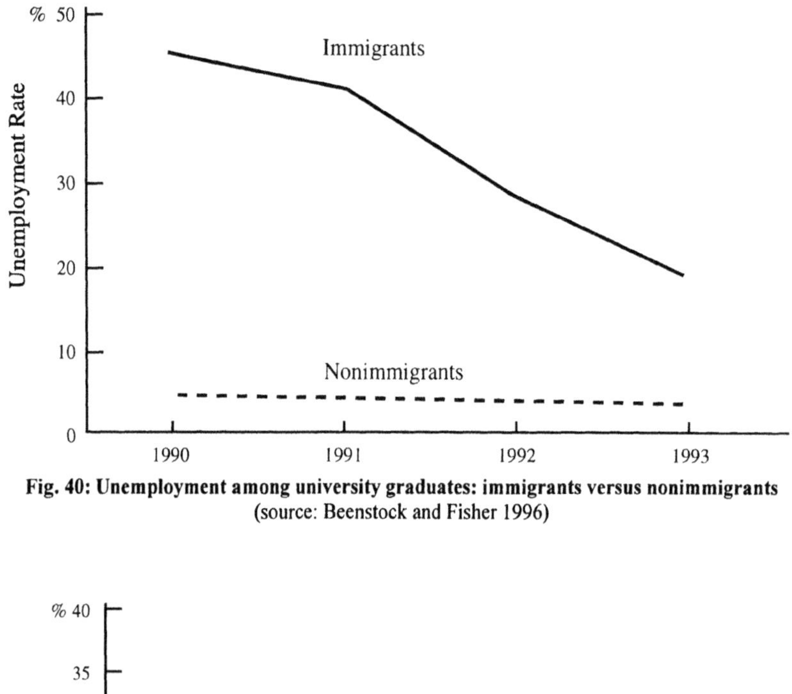

Fig. 40: Unemployment among university graduates: immigrants versus nonimmigrants
(source: Beenstock and Fisher 1996)

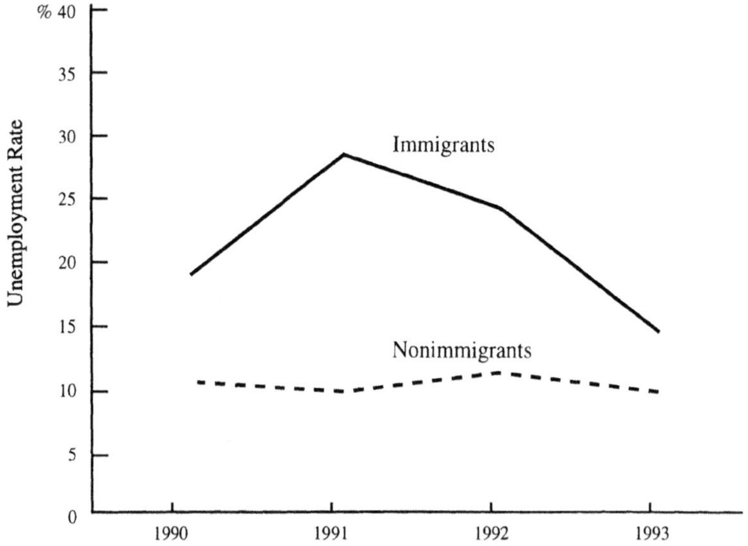

Fig. 41: Unemployment among non-university graduates: immigrants versus nonimmigrants
(source: Beenstock and Fisher 1996)

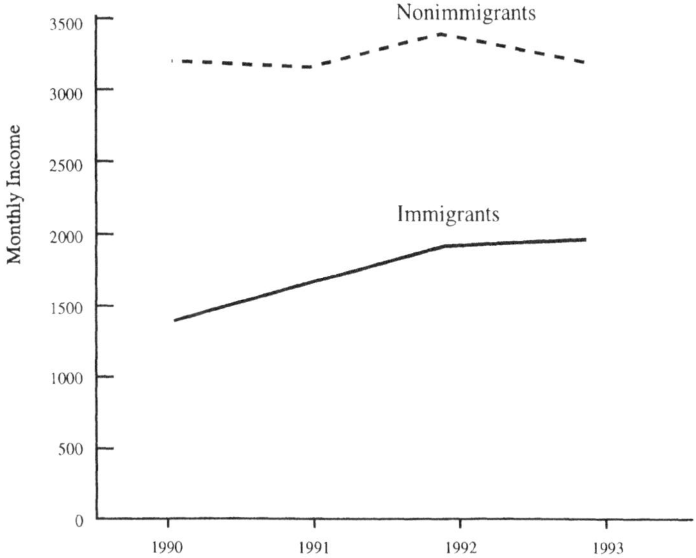

Fig. 42: Average monthly income: immigrants versus nonimmigrants
(source: Beenstock and Fisher 1996)

5.2 Economic Aspects of the Absorption of Immigrants from the Former Soviet Union in Israel: The Regional Level

One of the most conspicuous features of the immigrants from the former Soviet Union—as described in the previous section—is the large percentage of people with higher education and academic and scientific professions. The previous section considered the economic significance *on the national level* of the "import" of high-caliber human capital.

What geo-economic pattern has been produced by the immigration of high-caliber human capital? Are there differences in the quality of human capital between the immigrants who settled in the Israeli core and those who settled in the periphery? Has the socioeconomic profile of the immigrants in the periphery created a potential for a structural change in the economy of these regions? If so, has this potential been correctly utilized? We will discuss these questions in this section.

The differences between the immigrant and nonimmigrant populations in terms of rates of higher education and academic and scientific professions are particularly prominent in the peripheral districts of Israel (see table 13). In the Southern District, for example, approximately 65 percent of the immigrants in the labor force have scientific, academic, or technical professions, as opposed to only 30 percent of the nonimmigrants. Computations based on table 13 show clearly that in all districts the ratio of the percentage of immigrants with academic and scientific professions to the percentage of nonimmigrants with such professions is above 1, as is the ratio for higher education.

Table 13: Employment, Education, and Age of Immigrants and Nonimmigrants, by District

District	Scientific, academic, and technological occupations		13 or more years of education		Age cohorts					
					Immigrants			Nonimmigrants		
	Immigrants	Non-immigrants	Immigrants	Non-immigrants	0–14	15–64	65+	0–14	15–64	65+
Northern	58.6	29.6	48.5	25.0	24.1	60.0	15.1	33.8	58.6	7.6
Southern	65.2	30.9	50.2	24.3	22.3	65.4	12.3	33.2	60.6	6.2
Tel Aviv and Central	60.2	35.1	56.5	30.1	23.3	65.0	11.2	28.1	60.7	11.2
Haifa	60.8	37.4	54.6	30.3	20.8	68.0	11.1	26.1	60.7	13.2
Jerusalem	67.2	43.0	55.7	39.8	21.4	65.9	12.7	32.5	59.4	8.1
National average	60.4	35.2	53.1	29.9	22.6	65.0	12.4	29.7	60.2	10.1

The figures for the nonimmigrant population are an estimate based on the 1983 census. The figures for the immigrants are based on data from the Ministry of Immigrant Absorption.

These ratios are particularly high in the Southern and Northern districts. In other words, the immigrants' socioeconomic level in every district is higher than that of the non-immigrant population, and the differences are greatest in the periphery.

What this means is that the high-quality human capital has good potential for effecting structural change in the regional economy as well as the national economy. The main determinant of the location of high-tech industry is human capital. The level of human capital in the Israeli core was high even before the immigrants arrived, and therefore high-tech industrial parks opened there. The Israeli periphery has suffered for three decades from the departure of the young and educated (Lipshitz 1990). This migration perpetuated the traditional job structure and the lack of diversity in industry for many years (Gradus and Krakover 1976). The "import" of such highly developed human capital to the periphery produced a one-time locational factor supporting structural change in the regional economy—from a traditional structure to a modern technological structure. Although the central government and local governments in the periphery sponsored several projects to take advantage of the human capital, they were on a relatively small scale. As we shall see, most of the immigrants found jobs in the large traditional factories, and the opportunity for change in the regional economic structure—especially that of the peripheral regions—was lost.

We now take a more detailed look at the quality of human capital of the immigrants who settled in the development towns in the Galilee and Negev (the northern and southern peripheries), by addressing the following questions: Is the occupational distribution of immigrants in the development towns similar to that on the national level? Are the educational, occupational, and age distributions of immigrants in the development towns similar to the distributions of nonimmigrants in the development towns? In general, are there differences between the quality of human capital of the immigrants in development towns and the national average?

Tables 14 and 15 show that the situation of the development towns is worse than that of the country as a whole, but that the immigrants in these towns score higher for some variables than do the nonimmigrants in the same towns. In other words, the quality of human capital of the immigrants who settled in the development towns is lower than that of immigrants to Israel as a whole, but higher—and sometimes significantly so—than that of the nonimmigrant population of the development towns. Table 14 shows that in 1990 the percentage of immigrants with scientific and academic professions exceeded the national average in only 7 of 24 development towns; in the other 17 the figure was below the national average. In contrast, the percentage of immigrants whose occupations were in industrial production was above the national average in 20 of 24 development towns, and lower than the average in only 4 of them. In 1991 the situation was slightly worse from the development towns' perspective:

Table 15 shows that in that year only 5 of the towns had above-average percentages of immigrant scientists and academics (down from 7 in 1990), whereas 19 had below-average percentages. The distribution of production workers in 1991 was similar

Table 14: Distribution of Immigrants from the Former Soviet Union in the Labor Force by Occupation and Place of Residence, 1990

Development town	Scientific and academic	Liberal professions	Clerical	Production workers	No trade	Not stated	Total
Acre	17.02	18.97	6.76	11.76	33.64	11.85	100.00
Afula	23.57	16.86	5.71	9.54	33.58	10.75	100.00
Arad	26.36	20.16	3.88	8.53	31.01	10.08	100.00
Ashdod	18.72	18.94	6.12	10.00	35.41	10.82	100.00
Ashqelon	23.47	17.79	5.23	8.94	33.97	10.59	100.00
Beersheva	25.51	17.46	4.04	7.03	34.98	10.99	100.00
Beit She'an	14.49	14.84	7.77	10.60	40.64	11.66	100.00
Beit Shemesh	21.92	16.76	6.45	10.03	31.66	13.18	100.00
Dimona	21.92	16.76	6.45	10.03	31.66	13.18	100.00
Eilat	23.70	25.93	5.19	8.89	23.70	12.59	100.00
Karmiel	20.83	19.45	5.24	8.57	35.53	10.38	100.00
Ma'alot	20.97	20.59	3.96	9.72	34.02	10.74	100.00
Migdal Ha'emeq	18.14	17.64	7.27	10.94	34.72	11.29	100.00
Mitzpe Ramon	13.70	21.92	10.96	12.33	28.77	12.33	100.00
Ofaqim	17.97	25.78	3.91	9.38	29.69	13.28	100.00
Qiryat Gat	23.57	16.86	5.71	9.54	33.58	10.75	100.00
Qiryat Malakhi	28.16	16.63	4.00	6.12	35.06	10.04	100.00
Qiryat Shemona	19.35	22.48	5.21	10.20	29.32	13.44	100.00
Safed	36.55	12.56	3.36	3.36	32.66	11.51	100.00
Sederot	17.04	18.85	6.11	11.81	36.01	10.18	100.00
Shelomi	26.36	21.82	2.73	7.27	23.64	18.18	100.00
Tiberias	18.78	17.50	4.98	11.81	35.56	11.38	100.00
Upper Nazareth	17.43	20.10	6.54	10.00	35.84	10.10	100.00
Yeroham	10.64	17.02	6.38	34.04	19.15	12.77	100.00
National average	21.96	18.24	5.49	8.31	34.71	11.28	100.00

Source: Ministry of Immigrant Absorption.

Note: Beersheva, Ashdod, and Ashqelon are included in the table even though they are really no longer development towns.

Table 15: Distribution of Immigrants from the Former Soviet Union in the Labor Force by Occupation and Place of Residence, 1991 (partial data)

Development town	Scientific and academic	Liberal professions	Clerical	Production workers	No trade	Not stated	Total
Acre	17.45	18.12	7.11	10.74	34.23	12.35	100.00
Afula	22.15	16.89	5.25	10.27	36.99	8.45	100.00
Arad	20.74	17.22	5.80	9.31	37.08	9.84	100.00
Ashdod	16.76	18.83	6.81	10.18	35.82	11.60	100.00
Ashqelon	16.63	19.19	5.01	11.98	34.84	12.35	100.00
Beersheva	23.07	16.89	5.10	9.15	35.59	10.19	100.00
Beit She'an	12.76	12.07	6.90	18.62	34.14	15.52	100.00
Beit Shemesh	16.94	12.10	6.45	16.13	31.45	16.94	100.00
Dimona	13.11	25.00	6.71	11.28	31.71	12.20	100.00
Eilat	26.80	16.49	5.15	11.34	24.74	15.46	100.00
Karmiel	21.71	17.83	5.43	8.79	36.56	9.69	100.00
Ma'alot	20.49	15.57	5.33	13.11	37.70	7.79	100.00
Migdal Ha'emeq	13.85	18.21	4.62	14.62	38.46	10.26	100.00
Mitzpe Ramon	18.31	28.17	4.23	11.27	29.58	8.45	100.00
Ofaqim	18.95	18.42	5.26	12.63	28.95	15.79	100.00
Qiryat Gat	17.82	16.86	6.13	11.49	37.93	9.77	100.00
Qiryat Malakhi	12.94	16.42	6.97	16.92	32.84	13.93	100.00
Qiryat Shemona	21.43	15.18	3.57	11.61	39.73	8.48	100.00
Safed	28.47	16.55	3.38	4.63	33.99	12.99	100.00
Sederot	13.85	12.31	7.69	20.00	30.77	15.38	100.00
Shelomi	10.87	13.04	4.35	30.43	23.91	17.39	100.00
Tiberias	21.79	18.80	4.27	11.32	32.05	11.75	100.00
Upper Nazareth	15.47	20.76	5.59	12.67	34.33	11.18	100.00
Yeroham	10.00	20.00	10.00	20.00	32.50	7.50	100.00
National average	21.53	18.08	5.23	9.67	34.83	10.66	100.00

Source: Ministry of Immigrant Absorption.
Note: Beersheva, Ashdod, and Ashqelon are included in the table even though they are really no longer development towns.

to that of 1990 in terms of the number of towns, but the values were different: the distance from the national average was greater. In other words, production workers were more highly concentrated in the same number of development towns.

How do the immigrants compare to the nonimmigrant population in the Israeli periphery? A survey found that approximately 56 percent of immigrants in Qiryat Shemona are scientists, academics, or members of the liberal and technical professions. Of the immigrants to Dimona, 42 percent fall into these categories, and a similar figure prevails in Qiryat Gat (Lipshitz 1992). In 1983, scientists, academics, and members of the liberal and technical professions made up only 20 percent of the labor force in Qiryat Shemona and 23 percent in Dimona. About 32 percent of immigrants to Qiryat Gat are scientists and academics, and another 35 percent are members of the liberal and technical professions.[1] In 1983, only 20 percent of Qiryat Gat labor force consisted of

1. Although the data on nonimmigrants are based on the 1983 census, they do serve as an estimate. Even if

these professionals. Approximately 52 percent of the immigrants in Qiryat Gat have a higher education, compared to only 15 percent of nonimmigrants in the town. According to a report by the mayor's office in May 1990, about 60 doctors had moved to Dimona (a few who passed the licensing exam are employed in a night clinic), as had 200 engineers and 40 musicians. By June 1992 Dimona had taken in 100 nurses, 40 musicians, 90 doctors, 330 engineers, and 220 teachers (according to a report by the town's absorption coordinator). This figure, too, underscores the differences between the occupations of immigrants and nonimmigrants, and shows significant disparities in the level of higher education between immigrants and nonimmigrants in the development towns surveyed. The results of the surveys thus indicate that there is a substantial disparity between the quality of human capital of the immigrants and nonimmigrants in development towns.

Let us illustrate this pattern with a few examples from tables 14 and 15. In 1990, 21.96 percent of immigrants in Israel had worked as scientists and academics in the Soviet Union. In Arad and Safed the figures were 26.36 and 36.55 percent, respectively; whereas in Ofaqim it was 17.9 percent, in Yeroham 10.6 percent, in Upper Nazareth 17.4 percent, and in Beit She'an 14.4 percent. In the same year, 8.31 percent of the immigrants in the labor force had been industrial production workers in the USSR, as had 34 percent of those in Yeroham, 11.8 percent in Sederot, and 10.9 percent in Migdal Ha'emeq. In 1991, 21.53 percent of immigrants in Israel were scientists and academics, as were 26.8 percent in Eilat, 23.0 percent in Beersheva, and 28.4 percent in Safed, but only 13.8 percent in Dimona and 10.3 percent in Shelomi. In that year 9.67 percent of immigrants in Israel had been production workers in the USSR, as had 12.6 percent in Ofaqim, 10.9 percent in Qiryat Malakhi, and 16.1 percent in Beit Shemesh. These findings are corroborated by a study carried out in Sederot in August 1994, based on a census of immigrants in the town. This census found that 22 percent of immigrants in Sederot are scientists or academics, compared to 30 percent countrywide. Nineteen percent are members of the liberal and technical professions, compared to 34 percent countrywide, and 33 percent are industrial workers, compared to 15 percent countrywide (Lithwick and Noam 1995).

A comparative analysis (based on data not presented here) of the age distribution of immigrants in development towns and immigrants countrywide, as well as the age distribution of immigrants and nonimmigrants in the development towns, shows that the development towns have a relatively high concentration of elderly (above age 65) and older working-age (45–65) immigrants, and a relatively low concentration of the youngest age cohort (0–17).

The data show clearly that in only three development towns does the percentage of immigrants aged 17 and under exceed the percentage of nonimmigrants in this age group; in 21 towns the percentage of immigrants aged 17 and under is less than the percentage of nonimmigrants. In the elderly age cohort (65 and above), the picture is reversed: in 20 towns the percentage of elderly immigrants exceeds that of elderly nonimmigrants. In the working-age cohort there is a clear distinction between ages 18–44

---

we adjust the data for the nonimmigrants downward, there remains a significant gap between the immigrants and the nonimmigrants in these variables.

and ages 45–65: the percentage of immigrants in the older group—which has difficulty finding employment—exceeds the percentage of nonimmigrants in the same age group in almost all of the towns, whereas the percentage of immigrants aged 18–44 resembles the average for each town.

Data show that the youngest age cohort (0–17) is weakly represented in the development towns in comparison with the national average and that the elderly (65 and over) are fairly heavily represented. Immigrants aged 45–64—the older segment of the labor force—were found to be especially highly concentrated in these towns compared to the national average. For example, in 1992 the share of immigrants aged 0–17 was 13.6 percent.

In conclusion, the data presented in this section indicate that the peripheral districts and the development towns in these districts took in both human capital on a high professional and educational level and a relatively old population compared to the local nonimmigrant population and the overall immigrant population. These traits of the immigrants have immediate ramifications for employment (and unemployment) rates and for demand for social services (especially among the older people). In the next section we examine the geographical aspects of unemployment, with emphasis on the peripheral districts and the development towns.

5.3 Trends in Unemployment Rates over Time: The Regional Level

The unemployment rate in Israel, which increased steadily beginning in 1988 to a peak of 11.2 percent in 1992, fell in 1993 to an average of 10 percent, and in 1994 was only 7.8 percent (table 16). The increase in unemployment was substantial between 1988 and 1989 (from 6.4 percent to 8.9 percent). From 1988 to 1992 the Israeli unemployment rate rose more than 4 percentage points (table 16). The number of unemployed persons in Israel doubled between 1988 and 1992, from 99,900 to 207,600; in 1993 it fell to 195,000 (table 17). The increase in unemployment at the start of 1989 was caused by a slump in economic activity, due in part to the dramatic increase in wages in the wake of the stabilization plan (Bank of Israel 1994). After 1989 the continued increase in unemployment was chiefly due to a large increase in the supply of labor, as immigrants joined the labor force. Although the immigration also boosted demand and created jobs, the net result was a higher unemployment rate.

An additional 100,000 people were employed in 1993, and this produced a sharp decline in the unemployment rate. According to the Bank of Israel

Table 16: Unemployment Rates, 1988–1994, by District
(percent)

District	1988	1989	1990	1991	1992	1993	1994
Jerusalem	7.3	8.5	9.4	10.9	10.8	8.7	6.6
Northern	6.6	9.1	9.0	9.9	10.4	10.6	7.8
Haifa	7.0	9.5	10.7	13.0	13.5	12.0	8.5
Central	5.6	8.4	9.5	10.4	10.4	10.1	7.3
Tel Aviv	5.5	8.3	8.8	9.4	10.4	8.3	7.3
Southern	8.5	11.4	11.8	12.2	13.8	12.2	10.6
Total	6.4	8.9	9.6	10.6	11.2	10.0	7.8

Source: Central Bureau of Statistics, *Labour Force Surveys*, 1993, 1994, special publication series nos. 998 and 1024.

Table 17: Number of Unemployed Persons Aged 15 and Over, 1986–1994
(thousands)

District	1986	1987	1988	1989	1990	1991	1992	1993	1994
Jerusalem	12.1	10.9	12.4	15.0	17.0	19.9	20.6	16.7	13.5
Northern	19.8	14.2	15.0	21.7	21.9	26.0	30.1	32.5	23.8
Haifa	16.0	14.9	15.4	20.7	24.0	32.9	35.2	32.0	23.3
Central	19.5	17.8	19.9	30.1	35.7	42.8	44.0	44.3	34.1
Tel Aviv	22.6	18.9	21.9	33.7	36.8	41.2	47.7	39.2	35.2
Southern	13.5	12.7	14.0	19.7	20.5	23.0	28.0	27.7	27.2
Total	104.0	90.4	99.9	142.5	158.0	187.2	207.6	195.0	158.5

Source: Central Bureau of Statistics, *Labour Force Surveys*, 1993, 1994, special publication series nos. 998, 1024.

report, the increased employment was influenced by expectations of accelerated growth in the next few years and the quarantine of the territories, because of which workers from the territories were replaced with Israelis. Other factors included several policy measures designed to reduce unemployment, including subsidization of increased employment under the Business Sector Encouragement Law, which helped employers afford the adaptation and training costs involved in a substantial increase in employment; public-works projects; and a toughening of the criteria for obtaining unemployment compensation. Another factor had to do with the immigrants themselves. Many of them were determined to find work, and they began to seek jobs very soon after arriving in Israel (Lipshitz 1992).

A look at the integration of the immigrants in the job market reveals a mixed picture. Employment among immigrants rose substantially, and their unemployment rate dropped significantly; however, a large portion of the immigrants were forced to work in occupations that do not make use of their skills (Bank of Israel 1993). A survey conducted in May 1992 among a representative sample of 1,530 immigrants found that 42 percent of employed immigrants were working in their original occupation or in a related occupation. Most of the immigrants suffered a decline in their professional status upon moving to Israel: 62 percent worked in lower-prestige fields, and 59 percent worked in jobs that did not require their level of education (Bar-Tzur and Hendels 1993)

In 1993, unemployment rates in Israel fell for the first time since the start of the wave of immigration (table 16 and fig. 43). The unemployment rate decreased in all sectors—among nonimmigrants and immigrants and among men and women. Among the hard-core unemployed, however—those who had been out of work for more than a year—the rate increased. Many of these people are immigrants, especially older ones whose chances of finding work are slim. People in the 45+ age group, which includes a high proportion of women and the less educated, have had the hardest time finding jobs. Many of them (about 70 percent) have given up looking for work, because they do not believe they have a chance of finding any in Israel (Bar-Tzur and Hendels 1993).

The increase in unemployment is not geographically uniform. Unemployment rates remain high in the Northern and Southern peripheral districts, as well as in the Haifa District, where there is a particularly high concentration of immigrants (table 16 and fig. 43). For example, in 1988—just before the immigrants began to arrive—the national unemployment rate was 6.4 percent, while the rate in the Southern District was 8.5 percent and that in the Northern District was 6.6 percent. In 1992 the national unemployment rate was 11.2 percent, and the rates in the Northern, Haifa, and Southern districts were 10.4 percent, 13.5 percent, and 13.8 percent, respectively. It should be noted that the unemployment rates in the Tel Aviv, Jerusalem, and Central districts were below the national average. This spatial pattern changed slightly when the wave of immigration started (from 1990 on). Beginning in 1990 unemployment rates rose in the Central District, presumably because of the many immigrants who chose to settle there.

In many development towns in the peripheral districts, the unemployment rates are above the national and district averages. A graphic analysis of data not presented in this book shows a disparity every year between unemployment in the development towns and the national unemployment rate. However, this gap began to widen in 1989, the first year of the immigration wave. Whereas in 1988 the gap was 2.7 percent (the national unemployment rate was 6.4 percent and the average unemployment rate in the development towns was 9.1 percent), in 1989 it was 4.2 percent (8.9 versus 13.1 percent), in 1990 it was 4.1 percent (9.6 versus 13.7 percent), and in 1991 it was 4.8 percent (10.6 versus 15.4 percent). In 1992 the gap began to narrow; in 1993 and 1994 it was only 1.5 percent.

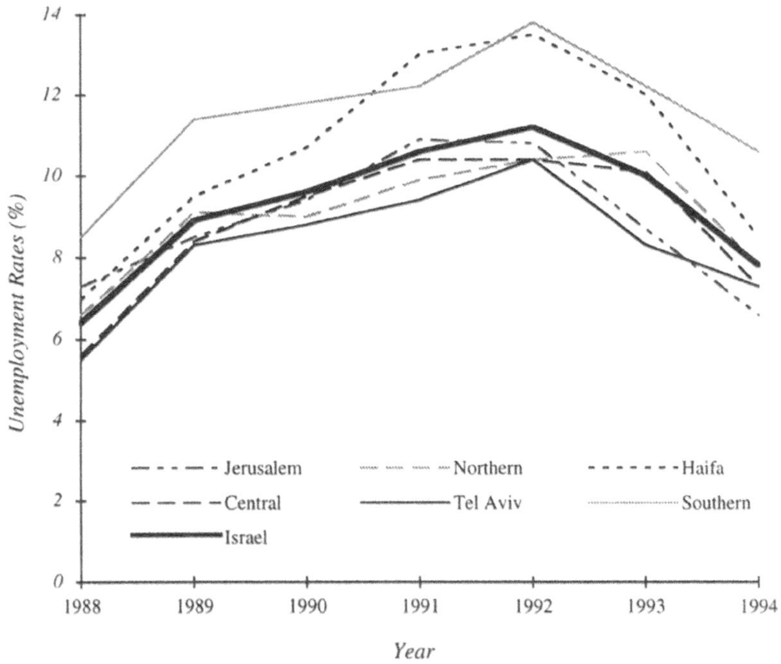

Fig. 43: Unemployment rates in Israel by district, 1988–1994

Several processes are pulling unemployment in peripheral development towns down to the national average. These include the integration of immigrants in jobs in the development towns, with downward adjustment of their occupations; the quarantine of the territories; an increase in jobs in development towns; migration of immigrants with higher education and academic or scientific professions from development towns to central Israel; a toughening of the criteria for obtaining unemployment compensation; and public-works projects.

The integration of the immigrants in the Israeli job market, inter alia in the peripheral districts, and the consequent decline in unemployment rates are themselves an impressive achievement, attesting to successful absorption in this domain. A more detailed evaluation of the degree of success should examine job satisfaction. Surveys conducted in five development towns in the peripheral districts indicate that approximately one-third of employed immigrants in Qiryat Shemona, 35 percent in Dimona, 46 percent in Qiryat Gat, 35 percent in Sederot, and 40 percent in Migdal Ha'emeq are not satisfied with their jobs (Lipshitz 1992). The surveys also indicate that a high percentage of immigrants (40 percent of all employed immigrants) continue to look for jobs while employed, and 40–50 percent of those employed expect to work in their present jobs for less than half a year. Another 15 percent expect to keep their present jobs for half a year to a year (Lipshitz 1992). These findings point to dissatisfaction and instability in the immigrant job market in the peripheral districts. Many immigrants would be willing to abandon the supportive "social and family network" for an "economic network" that provides a livelihood.

An evaluation of the success of the immigrants' integration in the job market must also look at the industries and kinds of factories in which they found work. Partial findings on this subject are available for the peripheral regions only. A survey conducted in development towns in the peripheral regions indicates that most of the immigrants have found jobs in *large, labor-intensive factories* (over 100 workers per factory). These are the "traditional" industries, such as textiles and food, which do not require a high level of education or a scientific or academic profession, and which pay low wages (Lipshitz 1992). Interestingly, factory owners who have hired immigrants state that the immigrants learn the job well and have a high work ethic, and they report satisfaction with the immigrants and an increase in worker productivity over time. The factory owners say that it took a few months for the immigrants to realize that they were not producing for the government, as was the case in the USSR. The factory owners prefer to hire immigrants who were production workers in the Soviet Union rather than university graduates who are willing to become industrial production workers. They have found that job stability is high among the former and low among the latter.

Small businesses in the peripheral regions took in relatively few immigrants. About 80 percent of small-business owners state that the immigration has not increased demand for their products substantially, and therefore they had no need for additional workers. The businesses that hired immigrants are chiefly those dealing in electrical appliances and furniture. These are essential products for immigrants, who received government assistance to purchase them (basic furniture, refrigerator, washing machine, oven, television). The immigrant workers thus served as intermediaries between the small-business owners and the immigrant consumers. Furthermore, the *rate of entre-*

preneurship among the immigrants (i.e., the percentage who open small businesses) is low. There are three reasons for this: lack of capital, lack of knowledge about local purchasing power and empty economic niches, and fear that the immigrants' purchasing power is low. Thus, even if they were to open "traditional" businesses, demand would be low (Lipshitz 1992).

The survey also indicates that the *public sector* hired few immigrants. This sector was "inflated" in the peripheral regions before the immigrants arrived, because the central and local governments had already expanded public-sector employment so as to reduce the high unemployment rates in these regions. The few public-sector jobs made available to immigrants were of two types: as "liaison workers" dealing with immigrants, such as clerks in welfare offices, and as "simple workers" doing such tasks as cleaning and guarding public places.

The success in lowering unemployment rates throughout Israel is a significant indication of the immigrants' integration. Is this integration long-term? Does the short-term drop in unemployment include many immigrants who are unhappy with their jobs and may ultimately join the ranks of the unemployed? More specifically, can we expect job stability among immigrants in the peripheral development towns?

Below we consider these issues with regard to the development towns in the Israeli periphery, focusing on the following questions: Are the immigrants employed in the development towns satisfied with their jobs? Do they regard their jobs as long-term or temporary? Surveys carried out in five development towns (Qiryat Gat, Qiryat Shemona, Migdal Ha'emeq, Dimona, and Sederot) indicate that businesses have not sufficiently taken advantage of the developed human capital that reached these towns, and that many of the employed immigrants are not working in the fields they studied in their countries of origin (Lipshitz 1992, 1997). The surveys also show that approximately one-third of the employed immigrants in Qiryat Shemona are dissatisfied with their work; the same is true of 35 percent of those in Dimona, 46 percent in Qiryat Gat, 35 percent in Sederot, and 40 percent in Migdal Ha'emeq. It should be noted that most of the employed immigrants in Dimona and Qiryat Shemona were production workers in the past, and therefore their satisfaction with their present jobs, which are also in manufacturing, is relatively high. In Qiryat Gat, in contrast, a high proportion of the immigrants employed in traditional industry are university graduates, and their satisfaction is fairly low. Countrywide, a study of immigrant job satisfaction found that approximately one-third are not satisfied (Naveh, Noam, and Benita 1993).

Many of the unemployed immigrants in Qiryat Shemona and Dimona have a higher education and scientific and academic professions, whereas the employed immigrants were production workers in their countries of origin. One of the reasons for the higher rate of job dissatisfaction in Qiryat Gat (approximately 65 percent) than in Dimona and Qiryat Shemona (25 percent and 33 percent, respectively) seems to involve differences in the immigrants' occupations in their countries of origin (professionals versus production workers). This factor may also explain the differences in rates of job-seeking while working: 66 percent in Qiryat Gat, and 40 percent in Dimona and Qiryat Shemona. Irrespective of the differences between towns, the high percentages who continue to look for work while employed says something about the instability of the immigrant job market in development towns.

The surveys carried out in the five development towns give an indication of the causes of this instability: a lack of correlation between occupation in the USSR and in Israel; employment in large traditional factories; and difficulties with the Hebrew language. About 70 percent of employed immigrants in Qiryat Shemona and 80 percent in Dimona work in traditional industries such as textiles, metals, and food. All of these pay low wages and do not require higher education or skills; consequently, immigrants—especially those with higher education and academic professions—are not interested in retaining their jobs.

One of the immigrants' most serious employment problems—especially in their first few years in the country—is lack of proficiency in Hebrew. Because they are worried about earning a living, they will take any job, even a simple one. Their failure to put enough time and effort into learning the language hurts them in the short and long terms. Because they do not know the language, they are unfamiliar with the supply of available jobs and do not understand the terms of employment. As a result, they have difficulty finding appropriate jobs, and some are even exploited. Knowledge of Hebrew is one of the most important conditions for job stability, for finding appropriate work, and for avoiding exploitation.

Job instability substantially increases the number of jobseekers in the development towns. In 1989 there were 1,055 jobseekers in Qiryat Gat; by September 1990 there were 1,285; and by May 1991 there were 1,775 (including 625 immigrants, 225 of them university graduates). In Dimona 620 immigrants were seeking work in July 1991; 345 of them were university graduates. The result was a 7:1 ratio of applicants to requests for workers. In 1991, the average number of jobseekers in Dimona was 1,250, whereas there were only 107 requests for workers. In May 1991, there were 865 jobseekers in Qiryat Shemona, 256 of them immigrants, but only 128 requests for workers were received that month by the Employment Service (based on an internal document by the Employment Service in Qiryat Gat).

5.4 Small Businesses as a Source of Jobs for Immigrants in the Periphery

One of the most important sources of jobs in the Western world consists of small and medium enterprises. The significance of small and medium enterprises in creating jobs has increased over time. Do small businesses have the potential to hire immigrants in the periphery? Surveys of small-business owners in Qiryat Gat, Qiryat Shemona, Dimona, Sederot, and Migdal Ha'emeq—all of them development towns in the Israeli periphery—found that hardly any new immigrants have been hired by businesses in neighborhood centers and in the center of town. About 90 percent of the business owners surveyed in Dimona and Qiryat Shemona and 95 percent of those in Migdal Ha'emeq said they had not hired a single immigrant and that, despite the immigration, their number of employees had not increased. A similar percentage said they were not interested in hiring immigrants. In Migdal Ha'emeq, too, the majority said they were not interested in hiring immigrants. The reason is that the immigrants budget carefully and purchase only essential products; therefore they have not boosted demand for products (other than groceries and the like), and business owners do not need additional workers. There is no need to hire immigrants as interpreters either, as the immigrants' purchasing power does not justify it. The owners say they would have hired immigrants as interpreters if there had been an influx of immigrant customers. A few immigrants were hired to work in real-estate agencies and electrical-appliance stores; their employers wanted to have someone to "mediate" between the company and immigrants. Of 32 businesses surveyed in Qiryat Shemona, only four had hired immigrants since the beginning of the present immigration wave: a grocery store, an electrical-appliance store, a furniture store, and a bank. Of 30 business owners surveyed in Dimona, only three had hired any immigrants: a grocery store, a furniture store, and an electrical-appliance store. Of 40 businesses surveyed in Migdal Ha'emeq, only two had hired any immigrants: a minimarket and a stationery/gift shop.

Most of the small-business owners do not expect to have any reason to hire immigrants in the near future. A few said they would be willing to hire an immigrant but do not need one. A few said that they had employed immigrants in the past but had had to fire them because of exaggerated wage demands and job instability, and that they would therefore be hesitant to hire immigrants in the future. Four of the 32 business owners interviewed in Qiryat Shemona said they would be willing to hire immigrants in the future: in a clothing and footwear store, a grocery store, a furniture store, and a gas distribution agency. Only three of the business owners in Dimona said they would like to hire an immigrant: two in the food industry and one in a furniture store. In Migdal Ha'emeq, only two of the 40 owners said they were interested in hiring immigrants. In both Migdal Ha'emeq and Qiryat Gat, electrical-appliance, furniture, and clothing stores are the most willing to hire immigrants. In both of the towns, however, only one immigrant was hired per business. The reason the businesses had no plans to hire additional workers, including immigrants, in the near future may be that the immigrants are not increasing demand for products substantially, and therefore they have not contributed to an improvement in the financial situation of the business. The owners of 24 businesses (75 percent of the Qiryat Shemona sample) reported that the immigration had not benefited them financially. Eight said that it had benefited them somewhat. Three of these own grocery stores and butcher shops in residential areas, four sell items that immigrants have to buy for themselves (large electrical appliances and furniture), and the eighth supplies steel doors to the Amidar housing company; immigration increased demand for these doors. Two of the 30 business owners in the Dimona sample reported that the immigration had improved their financial situation somewhat. In Migdal Ha'emeq only two out of 40 owners reported that the immigration had benefited them somewhat; one of them owns a minimarket. In Qiryat Shemona, Dimona, and Migdal Ha'emeq, not a single small business reported that the immigration had benefited it substantially. In the short term, most claimed that the immigration had not benefited them financially, and many even said that their routine business operations had suffered because the immigrants tend to enter and ask a lot of questions without buying.

When demand for products remains stable, businesses do not need additional workers. This is the main reason why small businesses in the development towns have not hired immigrants. Moreover, neighborhood businesses such as grocery stores and vegetable stores are concerned that the entry of immigrants into the neighborhood and the departure of nonimmigrants will burden them with the immigrants' consumption patterns. Although the survey of immigrants indicates that they tend to do most of their shopping locally (about 95 percent purchase products and services in town), the survey of businesses indicates that the purchasing volume is extremely low and focuses on basic commodities only. Conversations with immigrants reveal that fear of having no livelihood in the future leads them to save in the present.

It should be noted that in late 1995 a change occurred in the trend described above. In Dimona there are now signs of immigrants opening small businesses, and the percentage of immigrant workers employed by small businesses in the town has increased. According to the Dimona absorption coordinator, more immigrants work in garages, for graphics and sign companies, as drivers for private transport companies, and as sales personnel in stores. Recently an immigrant started a sewing shop in the town that employs about 50 other immigrants. Similar changes are taking place in other development towns as well, and they are clearly a change for the better, but they are still relatively insignificant compared to trends in the country as a whole. Approximately 28 percent of the immigrants are interested in setting up private businesses, and this percentage increases with age (Naveh, Noam, and Benita 1993).

5.5 Large Factories as Employers of Immigrants in the Periphery

The findings presented in section 5.4 indicate that small businesses have hired few immigrants. Have large factories in the periphery hired many? Are immigrants their work force of the future? Have the immigrants been taking nonimmigrants' jobs? These questions will be the subject of this section.

Surveys carried out in large factories in the same five peripheral development towns (Qiryat Gat, Dimona, Qiryat Shemona, Migdal Ha'emeq, and Sederot) found that these businesses have hired numerous immigrants. For example, in 1993 a textile plant in Migdal Ha'emeq employed about 500 workers, including 62 immigrants; in 1994, about 80 of 520 workers were recent immigrants. In a writing-implements factory, 70 of 120 workers in 1993 were immigrants, as were 87 of 140 workers in 1994. In a printed-circuits factory, 35 of 220 employees were immigrants, and in a shutter factory, 50 of 95 workers in 1994 were immigrants (based on absorption figures from the Migdal Ha'emeq municipality). Of 250 people employed by a Beit Shemesh electronics plant in 1993, 50 were immigrants, and of 50 workers employed by a food manufacturer, 10 were immigrants.

The factory owners report that the immigrants learn the work quickly and understand it well, and that their work ethic is high. Overall they are quite satisfied with the immigrants. They generally want to hire immigrants who were production workers in the USSR and not academics who have become production workers. Most of the factory owners report that immigrants replaced Arab workers—especially during the Gulf War—and the "me" generation, the children of the generation that founded the development towns.

All the factory owners report an absence of tension between immigrants and nonimmigrants, and in some of the plants nonimmigrants had initiated special activities to help and support the immigrants. This seems to be because hardly any local young people work in the large factories; this group, as we shall see below, is a source of social tension with the immigrants. The factories generally pay the minimum wage plus fringe benefits. Many factory owners in the development towns report that young natives work in the factories just long enough to become eligible for unemployment compensation. "After this period, they simply vanish," one owner said. The older nonimmigrants remain in the traditional factories; they do not feel threatened by the immigrants and are not worried about losing their jobs. This is apparently why no social tension has developed between nonimmigrants and immigrants in the factories. The fact that the young natives abandon their jobs as soon as they become eligible for unemployment compensation weakens the argument that they feel threatened by the immigrants. It may very well be, however, that some of the young people would consider remaining in the large factories if they were offered pay raises. It is hard to get a raise when immigrants are knocking on the door, willing to work for minimum wage, even for a short period of time.

The main complaint heard in the past about the immigrants is that their job skills are poor (i.e., their output is low), but the factory owners believe this is part of the system to which they became accustomed in the Soviet Union, and that in the long term their output will rise. Indeed, a limited survey carried out in factories in Beit Shemesh and Ma'alot in 1994 found that employers are satisfied with the immigrants' output. Hiring immigrants entails not only occupational retraining and the establishment of high-tech industries, but also a change in habits and work values; the newcomers must be taught how to work in a non-Communist economic regime. The factory owners believe that hiring immigrant scientists and academics for factory work is a short-term solution only. It is a steep price for these professionals to pay, and they tend to leave the factories after a short time. It is better to direct immigrants who were production workers in their countries of origin to traditional factories. Such workers are more satisfied with their jobs and are more likely to keep them.

Because of the importance of finding work for immigrants, particularly in development towns, we present below the remarks of factory owners in Migdal Ha'emeq, Qiryat Shemona, and Dimona. Their comments give a picture of various aspects of employing immigrants in their factories: the immigrants' skills, their job stability, recommendations for more successful employment of immigrants, the likelihood of hiring workers in the future, and so on. Although the survey was carried out in 1992, the findings appear to be important even today.

Qiryat Shemona

A textile plant with about 200 employees took in 20 Soviet immigrants. According to the plant manager, all the Russian workers who apply for jobs say they are engineers, and this is hard to believe. Not a single one of them had any background in textiles. They start to work without asking questions, but soon, when they discover that the work is simple and not appropriate for them, they start to leave. All the immigrants hired work in production; none works in planning.

The plant manager reported satisfaction with the immigrants, and in his opinion their output is similar to that of long-time Qiryat Shemona residents. The immigrants have not boosted output, but they have not hurt it, either. They work in all departments;

the factory has no "Russian unit." The immigrants stick together and have not yet developed social relationships with the other workers. One reason for this is that they do not know Hebrew. The immigrants earn the minimum wage plus fringe benefits. At this point there are no plans to expand the factory, as output depends mainly on worldwide demand and competition is sharp. Therefore there are no plans to hire additional workers. If more immigrants are hired, it will probably be at the expense of the immigrants already employed, nonimmigrant residents of Qiryat Shemona, or Lebanese workers (whose output is much higher than the average). If and when the business is expanded, immigrants may be hired, because the young people in Qiryat Shemona are not interested in working in textiles. The immigrants hired will be experienced production workers and not university graduates; nevertheless, all new employees will take an on-the-job training course. "We won't take a food engineer to pack socks," the plant manager said.

A factory that produces disposable medical equipment (blood-transfusion equipment, etc.) employs about 220 workers, including 50 immigrants. The company seeks production workers as well as professional workers (e.g., production-line managers and quality-control personnel). The hiring process is meticulous, because the final product involves human life and the work therefore entails great responsibility. The immigrants are afraid to take management positions, mainly because of their lack of proficiency in the language and because they are not used to working in Israel. The factory's employment capacity depends, of course, on demand for its products; much of its output is intended for export.

Immigrants have already been hired to work in all parts of the factory: engineering, planning, and production. Turnover among immigrants is low compared to turnover among nonimmigrant workers. The immigrants have a high work ethic, they are strongly motivated, and their output is above average. There is hardly any tension between immigrants and nonimmigrants, and social ties are starting to develop between them. The firm itself deliberately promotes social ties between nonimmigrants and immigrants. For example, it invited all of its employees to a social weekend at a hotel. The company, in conjunction with the Jewish Agency and the Ministry of Education, intends to offer technical courses at the plant, as well as English and Hebrew courses.

As in the other factories, the starting wage of production workers is the minimum wage (NIS 1,160). With time, employees receive raises and increased fringe benefits. Immigrants who work as engineers earn a starting salary of about NIS 1,800. The company believes in investing in its employees so as to make them happy and convince them to stay. Anyone who wants to work there can get a job: "If a professor shows a genuine, honest inclination to work in quality control, he will be hired, but if a laboratory technician who is ostensibly qualified shows signs of hesitation, he will not be hired," said the plant's personnel manager.

A factory that produces clothing for the military has approximately 100 employees, 18 of them immigrants. At one point about 150 people worked there, including 25 immigrants. The downsizing was due to a decrease in orders for the company's products. The plant needs mainly unskilled workers, and most of its recent additions have been Russian immigrants. According to the plant manager, if the company expands in the future about half of the workers hired will be from the former Soviet Union. The factory owner said he was very satisfied with the immigrants' work ethic, and—like the owners of the other factories—he is afraid that the Israelis will eventually ruin the immigrants.

Turnover among the immigrants at this factory is high, partly because the Russians are prepared to take any job until they discover that it does not suit their skills and qualifications. For instance, mining engineers and mechanical engineers who began to work as cutters (after having said they had a suitable background) soon quit or were fired. Naturally, this causes distrust of the documentation that the immigrants present to prove their occupational backgrounds and skills. Those who were production workers, especially tailors, in the USSR exhibit good professional ability. Others are poorly qualified, and time must be invested in guidance and vocational retraining. The immigrants are still not aware of the importance of product quality. Leather cutters who worked as cutters in this factory discovered that precision is crucial in cutting fabric, unlike in cutting leather. The immigrants were not used to the quality control conducted in the factory, and sometimes they had to redo their work in order to meet the required standards.

Dimona

An electronics factory employs about 360 workers, including 95 immigrants; 14 of the immigrants were engineers in the USSR. Most of the immigrants are production workers here. The company plans to expand soon because of increased demand for its products, and most of the workers hired will probably be immigrants. Preference will be given to those who were production workers, electronics engineering technologists, or electronics engineers in the USSR (here the engineers will work as engineering technologists or production workers).

As the factory expanded over the years, it needed more and more technicians and engineering technologists. Qualified young workers were hard to find in Dimona and Yeroham because they had moved to Beersheva and the central region. Even among students in Beersheva training to become technicians and engineering technologists, it was difficult to find suitable candidates, because most of them had been sent by a particular firm to study and were supposed to return there when they completed their studies. In the past, there were problems with production workers, too, some of whom chose to work for 151 days and then leave the company and collect unemployment compensation. For now, the immigrants from the USSR fulfill two functions: they work as technicians and engineering technologists and they are disciplined production workers. Therefore, the new workers from the USSR should not be seen as supplanting nonimmigrants, despite allegations by the nonimmigrants; nevertheless, there is sometimes friction on these grounds between nonimmigrants and immigrants. The immigrants' job stability is fairly high for three reasons: the screening method, the company's policy of taking care of financial and social matters unrelated to the immigrants' jobs, and the company's policy of letting the immigrants know that they are wanted and that they contribute to the factory's output.

A textile plant employs about 650 workers during its high season. Of these, about 75 are immigrants. There is high turnover among the immigrants: of 150 who used to work there, about 75 remain. Fifty have worked there for five months already, and the others are newer. Most of the turnover is among university graduates: an economist and a musician were employed as production workers and had to leave after a short time. There is also turnover among those who were production workers in the Soviet Union, especially because they are "grabbed" by contractors who pay them NIS 200 more. The immigrants do not understand that this hurts them because the contractors rarely pay National Insurance and other similar payments (the immigrants in the factory earn the minimum wage, and this already includes incentive pay and fringe benefits). All the workers are employed as production workers or in the basic services. In the entire factory, there is only one university graduate (a doctor of chemistry) who is expected to work in his field in the future. In the short term, he will serve as the department technologist.

The factory owner reports that the immigrants' output is low (at least initially). Nevertheless, it can be assumed that in the future these immigrants will constitute the bulk of the factory's personnel because of the refusal of young people from Dimona and the surrounding area to do this work and because of difficulty employing workers from the territories. For these reasons, the immigrants should not be considered a threat to the jobs of development-town residents.

Factories in which the immigrants' output is low face a difficult dilemma: "economics versus Zionism," as the plant manager put it. For economic reasons it is best to hire Arabs from the West Bank, but for reasons of Zionism it is best to hire immigrants. The factory owner said he does not sense any tension between immigrants and nonimmigrants, even though no ties have been formed between them. When a shift ends, the nonimmigrants tend to go straight home, whereas the immigrants "zigzag," i.e., they stop repeatedly to talk to other immigrants.

A futuristic technology center has been in operation in the Dimona area for about a year. Its aim is to run a "technological incubator" for the development of marketable products. The idea of the incubator is to help entrepreneurs develop a product that they will be able to market within two years. About half the members of the product-development team must be immigrants. In the past year a professional committee approved five products, and 25 immigrants, including high-level university graduates with degrees in applied mathematics, chemistry, and computer science, are working on them. Another five products are being reviewed. If they are approved, 25 to 30 more immigrant university graduates may be hired. Employees earn about NIS 2,500 a month, and this sum could increase substantially if the products succeed. In such a case the developers would receive a certain percentage of the value of the products sold.

Migdal Ha'emeq

A factory that produces printed circuits for the electronics industry employs about 200 workers in three shifts; about 30 of them are immigrants, and their turnover is low. At first, the immigrants hired did not know enough Hebrew. Because of the resultant difficulties, management decided not to hire immigrants who did not have a reasonable knowledge of Hebrew. The immigrants at the plant were hired to work in all aspects of production: as laboratory technicians, technicians, and engineering technologists. Hiring decisions are based on whether the applicant has the required skills; no special preference is given to immigrants. The starting wage of production workers is the minimum wage, but promotions are given. The factory owners have learned not to hire university graduates as production workers. Not only do they lack experience in factory work, but their job stability is low.

An electronics and computer factory that employs 500 workers has hired about 40 immigrants. The immigrants came to the company in various ways. The factory has been seeking workers for several years, and most of those hired have been immigrants. Although this is a high-tech company, the unit in Migdal Ha'emeq deals with production; research and development is done in the central region. Therefore, the demand in Migdal Ha'emeq is mainly for production workers on various levels, and a university education is not required. Immigrants have been hired as production workers, and their job stability is fairly high. They are motivated to work because of the company's good image, and as a result they have a high work ethic. The factory owners believe that the immigrants' job stability will remain higher than that of the young native Israelis completing their military service. Wages are among the best in the market and, in keeping with the American method, depend on workers' output.

A factory that produces writing implements (pens of various sorts) employs about 100 workers, 80 percent of them recent immigrants. The factory moved from Haifa to Migdal Ha'emeq in order to expand. When it started hiring immigrants, turnover was high. Some of the workers left of their accord and others were fired. In many cases immigrants quit without giving advance notice, and this harmed management's confidence in them. The situation has now stabilized somewhat, and the immigrant workers' output has increased over time. Here, too, the factory owner raised the issue of employing university graduates. Many of the university graduates will have to remain production workers. The owner has a favorable attitude toward immigrants, and he will try to help them integrate, but ultimately economics will determine whether they retain their jobs. At first the immigrants went directly to the factory without having been sent there. Later, immigrants referred each other to the plant. Because they felt a certain obligation to the company, they did their own pre-screening of their immigrant friends who wanted to work there. An expansion of the business will create a need for additional workers, and the owner believes that many of these workers will be new immigrants.

Thus, the surveys of factory owners are optimistic regarding the employment of immigrants. This optimism is lowest with respect to highly educated immigrants with academic and scientific professions. In general, the owners are satisfied with the immigrants' work ethic and output. Their satisfaction is greatest with regard to those who were production workers in the USSR and lower (sometimes substantially so) with regard to those with academic and scientific backgrounds. The factory owners regard the new immigrants as their potential pool of employees for when their businesses expand and demand for workers rises.

CHAPTER 6

CONCLUSION

Israel is perhaps the only country in the world whose immigration policy has hardly anything to do with quotas. Israel is the state of the Jews, as asserted in its Declaration of Independence, read out by the first prime minister, David Ben-Gurion, when he proclaimed the establishment of the State in May 1948. Consequently, under the Law of Return, the country's gates are open to all Jews in the world. Israeli governments in the past and present have not only encouraged Jewish immigration, but have even taken care of the immigrants in their first years in Israel, helping them obtain (or in some cases giving them) housing, employment, education, and welfare. Whereas few—if any—other countries that take in immigrants pay attention to the geographical dispersion of the immigrants, this dispersion, which leads to geographical dispersion of jobs and infrastructure, is an important aspect of immigrant-absorption policy in Israel.

Israel is one of the youngest states in the world (it will celebrate its 50th birthday in 1998) in which immigration has had such a decisive impact on the settlement network, on the spatial distribution of the population, and on the spatial distribution of economic activity. *Thus, the present book explores one of the most important factors, if not the most important one, in determining the economic, settlement-related, social, and cultural geography of Israel.* When the immigration factor was weaker (in the 1960s, 1970s, and 1980s), a different population flow shaped the settlement and economic geography of Israel: internal migration. The book examines this flow, too, which is affected directly and indirectly by the spatial distribution of immigration to Israel before and after independence.

As the book indicates, there is a high spatial correlation between the geographical distribution of the affluent population in Israel and that of the pre-State immigrants, and there is a high spatial correlation between the geographical distribution of the disadvantaged population in Israel and that of the immigrants of the 1950s. The spatial encounter between these two waves of immigration is what created the regional economic and social disparities in Israel; in other words, it determined the spatial-socio-economic structure of core and periphery. The flow of internal migration, which gained momentum beginning in the 1960s, is mainly from the periphery to the core.

In general—as is explained in detail in the book—until the wave of immigration of the 1950s Israeli society was relatively uniform geographically. The pre-1948 immigrants from eastern and western Europe were on a relatively high social and cultural level. For ideological reasons, many of them (though not the majority) had turned to rural settlements in regions that would eventually be referred to as the periphery. Many of the top political and military leaders and leading authors and poets came from the kibbutzim, moshavim, and moshavot in these regions. In other words, when Israel gained its independence, *its geographical periphery was by no means a social or economic periphery* (according to the socioeconomic traits of the population). It *was* a demographic periphery: about 80 percent of the population (most of it made up of immigrants who arrived in 1882–1948) was concentrated in what would become the Tel Aviv, Central, and Haifa districts, and 8 percent lived in the north and south (only one percent of the Jewish population lived in the south). The reasons for this are explored in chapter 2.

When Israel gained its independence, its Jewish population numbered 716,000; within the next ten years one million immigrants arrived, most of them from North Africa and Asia. The newcomers came with no financial capital and little education, and few were professionals. As the book describes, Israeli governments since 1948 have implemented spatial policies aimed at dispersing the population from the core to the periphery. For geopolitical, geosecurity, and geo-economic reasons, individuals and entrepreneurs were and are granted government incentives to migrate to and/or open businesses in the peripheral regions.

As it turned out, Israel's population-dispersion policy was a policy for dispersing the immigrants of the 1950s. Relatively few "nonimmigrants" (pre-1948 immigrants and their children) answered the government's call to move to the Negev and develop the barren land. Most preferred to remain in the central region; indeed, internal migration in Israel is from the periphery to the core, not from the core to the periphery. Immigrants (and their children) who initially settled in the peripheral regions and improved their economic situation tend to leave their area of residence and migrate to the core. They are joined by educated, talented young people who grew up in development towns in the periphery and seek work in the core that suits their skills and qualifications. Internal migration thereby hinders the efforts of the peripheral regions to effect a structural change in their economy from a traditional one to a modern one that requires educated and skilled human capital.

Thus, in Israel "planning from above" (by the government) works geographically from the core to the periphery, and "planning from below" (by market forces, which are manifested in internal migration) works from the periphery to the core. In general, the spatial results indicate that "planning from below" perpetuates "planning from above." In the early 1960s, when immigration to Israel decreased dramatically and internal migration became the dominant factor shaping Israeli spatial patterns, *the process of population dispersion in Israel ceased on the national level, and an accelerated process of population deconcentration began on the metropolitan level.* Affluent and young families migrated from the central cities of the Tel Aviv and Haifa metropolitan areas to the towns in the outer and distant rings. Young families also moved to community settlements in the Galilee and in Judea-Samaria (the West Bank). These localities, with high-quality housing and attractive community conditions, were founded by the Israeli government in Judea-Samaria beginning in 1967 (the Six-Day War) and in the Galilee beginning in the mid-1970s. Many of the localities in western Samaria became the distant suburbs of Tel Aviv, and many of those in the Galilee became distant suburbs of Haifa. A similar process is taking place in the Jerusalem metropolitan area, too. The reasons for migration to the community settlements in Judea and Samaria are ideological-religious (settlement of all parts of the Land of Israel, particularly the West Bank), and socioeconomic (the classic considerations of migrants to the suburbs). Internal migration in Israel thus promotes concentration on the national level and deconcentration on the metropolitan level.

This book has hardly mentioned internal migration by Israeli Arabs, because its subject is migration on the geographical level of core and periphery, whereas the Israeli Arab population migrates mainly *within* the peripheral regions, where it has lived for many generations. In summary, Israeli Arabs, who are concentrated in the peripheral regions, migrate mainly within the peripheral regions: between Arab localities and from the center of the Arab town/village to its periphery. Although some Arabs migrate from the Galilee to the city of Haifa, very few do so. Thus we see that the Arab population, like other minorities in the world, prefers to remain and live in its own "cultural space." In addition, with the rise in the motorization rate among the Arab population and the improvement in the road infrastructure from the Galilee to Haifa and Tel Aviv, many Arab localities have become distant suburbs of Haifa, and many of their residents commute to the employment fields of this metropolitan area (as well as to the Tel Aviv metropolitan area).

Internal migration in Israel might have remained the only player on the stage of settlement and economic geography in Israel for a long time, if there had not been an additional "noise" on the country's population landscape—a mass wave of immigration from the former Soviet Union. These immigrants have been coming to Israel since late 1989, and by 1996 they numbered over 700,000, equal to approximately 15 percent of the Israeli population.

From a geographical perspective, this wave of immigration, unlike the wave of the 1950s, did not substantially alter the spatial distribution of the Israeli population. The share of immigrants in each district as a percentage of the total immigrant population resembles the share of the nonimmigrant population by district as a percentage of the Israeli population. In other words, on the national level there was hardly any change in the distribution of the population, but significant changes occurred on the level of individual towns and sections thereof. Because this book is based on "large" geographical units (districts, subdistricts, core/periphery), it pays no attention to changes on the intra-urban level and only briefly mentions changes on the level of localities. The immigrants in the 1990s chose to settle in neighborhoods on a low socioeconomic level. In the metropolitan areas they "revived" neighborhoods in physical and demographic decline. Immigrants also settled in some peripheral development towns, where many apartments were empty due to out-migration in the past few decades.

This wave expedited deconcentration processes on the metropolitan level. In towns in the periphery of the Tel Aviv and Haifa metropolitan areas where there were large land reserves, an accelerated process of residential building began, and their populations have increased significantly in the past five years. In addition, the development towns (founded in the 1950s to absorb the immigrants who were arriving then) that are situated geographically on the "seam" between the periphery of the core and the national periphery have grown more rapidly than the national average due to the recent immigration. These towns enjoy two advantages of geographical location: they are in the commuting range (slightly over the average Israeli commuting time) of the employment fields of the metropolitan areas, but because they belong administratively to the peripheral districts, residents and entrepreneurs enjoy government support and benefits. The immigration from the former Soviet Union thus accelerated the expansion of the Israeli metropolitan areas toward regions that used to be part of the national periphery.

Although on the macro (district) level there were no substantial changes in the population distribution, on the micro (local) level dramatic changes took place in several localities, most of them situated in the peripheral regions of Israel (the "seam" localities). Although in absolute numbers the cities in the core took in more immigrants than did the peripheral towns, in relative terms immigrants constitute a low percentage of the population of the core and a relatively high percentage in some peripheral development towns. In some development towns immigrants constitute more than 30 percent of the population. These towns, also known as "new towns," have seen major changes in their built-up area (residential, commercial, and industrial). The accelerated process of renewal that these towns experienced has, of course, had a direct impact on residents' reactions to the absorption of immigrants in their localities. Nonimmigrants in these towns fear that the immigrants will take away their jobs; this explains their opposition to immigrants' moving in. Interestingly, the owners of large factories in the peripheral development towns reject the idea that immigrants will take jobs away from long-time residents. As they see it, the immigrants have taken factory jobs that nonimmigrants did not and do not want. In the major cities, where immigrants account for a low percentage of the population, the "threat" to the nonimmigrants is hardly felt, and therefore attitudes toward the immigrants are more favorable.

The wave of immigration from the former Soviet Union brought a high percentage of people with higher education and academic and scientific professions. Even if we assume that some of the immigrants' reports are inaccurate and that their qualifications sometimes fall short of Israeli standards, Israel received "ready-made" what other countries seek to obtain through selective immigration policies that they have recently adopted based on educational and occupational parameters. However, the Israeli economy was not prepared to make a structural change in its economy in such a short period of time, and therefore doctors, scientists, and academics have had to adjust their occupations downward. The result was a loss of billions of dollars to the Israeli economy.

Nevertheless, the economic absorption of the Soviet immigrants in Israel can be regarded as a success story. Within a very short period of time, unemployment rates among immigrants lessened (more rapidly among non-university graduates than among university graduates), and the gap between the unemployment rates of immigrants and nonimmigrants is narrowing substantially. Among the younger generation, the disparities between immigrants and natives are negligible.

From a geographical standpoint, immigration from the former Soviet Union has not narrowed the gap between the unemployment rates in the core and peripheral districts (as a matter of fact, the gap widened slightly). The peripheral districts took in a high proportion of immigrants with a high educational level and academic and scientific professions. The percentage of immigrants in these categories in the peripheral districts was more than twice that of the nonimmigrant population. Due to the traditional employment structure in these regions, a disparity developed there—more than in any other part of the country—between the supply of educated and skilled personnel and the demand by factories and businesses. Therefore, the peripheral regions have a relatively high percentage of unemployed university graduates. Many of them have sought jobs as factory workers, but they gave up quickly, as did the factory owners ("We prefer those who were production workers in the former Soviet Union and not professors").

It should be stressed that the "absorption package" that the immigrants received gave them "breathing space" of a few months to a year, without the burden of having to find work and earn a living. During this time, a profound structural change could have begun in the economy of the peripheral regions in the direction of high-technology industry, which relies mainly on educated, skilled human capital. Although several important actions were taken in this direction, they were too little and, in a certain sense, too late. Although more immigrants migrated from where they had initially settled in the core regions to the periphery than in the opposite direction, they moved mainly to the "seam localities" between the national periphery and the periphery of the metropolitan areas. Furthermore, it has been hypothesized that this migration is taking "strong immigrants" from the periphery to the core for job-related reasons and "weak immigrants" from the core to the periphery due to housing considerations.

The Israeli core regions, which before the current wave of immigration had begun to make a structural change in their economy in the direction of accelerated development of the business sector (financial, computer, and scientific) and high-technology industries and had started to flourish economically, took in the immigrants with no unusual shocks in their job markets. The immigrants merely accelerated the economic growth of the core and the expansion of its built-up area.

Thus, as the twentieth century nears its end, Israel is characterized by two core regions, which include areas that once belonged to the national periphery, and a shrinking national periphery. To describe this, some people today call Israel "the State of Tel Aviv." This designation may be a bit humorous, but there is no doubt that the boundaries of the core and peripheral regions on which this book is based no longer reflect the spatial processes that have taken place in the 1990s. Macro-level sociospatial research in Israel (core versus periphery) at the end of this century, and certainly in the next century, will have to use regions whose boundaries are different from those described in this book. In particular, researchers will have to address the tremendous expansion of the core regions of Israel.

The book does not deal with immigrants from Ethiopia, who arrived in Israel in two impressive operations: Operation Moses (1984–85), which brought about 8,500 immigrants; and Operation Solomon (1991), which brought more than 14,000 immigrants. The spatial distribution of the Ethiopian immigrants in Israel, the absorption policy adopted for them by the Israeli government, their cultural and social traits, their integration in Israeli social and economic life, family relations, and the sizable gaps between the parents and children require a study with an anthropological-sociological focus; the geographical aspect is less important.

The book also does not address the issue of foreign workers, which is a fairly new phenomenon in Israel (about five years old). Officially, they number about 200,000, but unofficially their number is estimated at 350,000. The demand for foreign workers began when, for security reasons, closures were imposed on the West Bank and the Gaza Strip, thereby keeping thousands of Palestinian workers out of Israel. For the most part, the foreign workers replaced Palestinians: Thais in agriculture and Romanians in construction. Filipinos have provided chronic-care services for many years, irrespective of the closures between Israel and the territory of the Palestinian Authority. As in other countries, many of the foreign workers live in the major cities: Tel Aviv, Haifa, and Jerusalem. In Tel Aviv they already have their own colony in poor areas near the old central bus station, with schools, mosques, and churches for the foreign workers.

Immigration to Israel and internal migration between Israeli regions are one element in the global patterns of migration between and within countries. The geographical, demographic, economic, and political characteristics of these flows of migration were described in chapter 1. It turns out that in some ways migration to and within Israel resembles that in other countries, whereas in other aspects it differs.

As we already mentioned, immigration quotas, which occupy all other countries that take in immigrants, are not an issue in Israel's immigrant-absorption policy. Israel is open to all Jews, irrespective of age, financial situation, and state of health (with very few exceptions). On the other hand, in most countries the geographical dispersion of immigrants is left to market forces, and policies to guide it are nonexistent; in contrast, Israel—in the past even more than in the present—invests extensive resources in dispersing the immigrants geographically toward the periphery. In the past this was done for geopolitical, geosecurity, and geo-economic reasons; now planning factors are also relevant.

For economic reasons, immigrants the world over tend to concentrate in the core regions (metropolitan areas) and in relatively poor neighborhoods in these regions. In the 1950s most immigrants to Israel were directed by the government to new towns built in the national peripheral regions and in new neighborhoods built in the periphery of the small and medium-sized towns. Due to a combination of the immigrants' socioeconomic traits—described in chapter 2—and the geographical-dispersion policy, a socioeconomic periphery was created both on the national level and on the urban level (poor neighborhoods).

Internal migration in Israel resembles that in Western countries: it moves from the periphery to the core and to its periphery (interregional migration) and from the core to its periphery (suburbanization). In Israel and in other countries, this migration expands the boundaries of the core, and simultaneously increases the commuting range to the core on the national and regional levels.

As in other countries, the Israeli core suffered a decline in the 1970s and 1980s because the periphery of the core was flourishing; in the 1990s, in contrast, the first signs of renewed growth of the core regions can be seen. These processes, which are already under way in some Western countries, are a manifestation of structural changes in the global economy and of the re-emergence of agglomeration economies with respect to locational considerations of businesses.

Immigration from the former Soviet Union in the 1990s, which put an end to internal migration's distinction as the only demographic factor shaping spatial patterns in Israel, reinforced the spatial patterns fashioned by internal migration in the 1960s, 1970s, and 1980s. The impact of this immigration wave on the geographical distribution of the Israeli population and on its economic and geo-economic development resembles the impact of waves of immigration to Western countries. However, while Western countries must use selective policies in order to receive a high proportion of immigrants with higher education and scientific and academic professions—which are needed for the development of business and high-tech industry—Israel received these ready-made.

If Israel had been able to gear up appropriately to absorb this human capital economically, we would have seen higher rates of growth in the past few years, as well as significant structural changes in the regional economy, especially the economy of the peripheral regions. Nevertheless, the rapid integration of the immigrants from the former Soviet Union in the Israeli economy is a success story, even in an international comparison.

BIBLIOGRAPHY

Addleton, J. 1991. The impact of the Gulf War on migration and remittances in Asia and the Middle East. *International Migration Review* 29:509–526.

Amiran, D., and A. Shachar. 1971. The Tel Aviv conurbation. In *The Land of Israel: Studies of the Land and Its Antiquities*, vol. 10, 198–217. Jerusalem: Zalman Shazar (Hebrew).

Appleyard, R. T. 1989. Migration and development: Myths and reality. *International Migration Review* 23:457–485.

Arin, C. 1991. The housing market and housing policies for the migrant labor population in West Berlin. In *Urban Housing Segregation of Minorities in Western Europe and the United States*, E. Huttmann et al., eds., 199–214. London: Duke University Press.

Bailey, T. 1985. The influence of legal status on the labor market impact of immigration. *International Migration Review* 19:220–238.

Bank of Israel. 1990. *Economic policy in a time of immigration*. Jerusalem: Bank of Israel Research Department (Hebrew)

Bank of Israel. 1994. *Bank of Israel Report 1994*. Jerusalem: Bank of Israel (Hebrew).

Bank of Israel. 1995. *Bank of Israel Report 1995*. Jerusalem: Bank of Israel (Hebrew).

Bar, A. 1990. Absorption of new immigrants as an impetus for the development of localities. In Gonen 1990, 32–33 (Hebrew).

Bar-El, R. 1985. Industrial dispersion as an instrument for the achievement of development. *Economic Geography* 61:205–222.

Bar-Tzur, R., and S. Hendels. 1993. *Needs and preferences of immigrants from the former Soviet Union with respect to health and employment*. Tel Aviv: Institute for Economic and Social Research, General Federation of Labour in Israel (Hebrew).

Bar-Yosef, R., and D. Padan. 1976. The oriental communities in class stratification in Israel. In *The Social Structure of Israel*, S. N. Eisenstadt, H. Adler, R. Bar-Yosef, and R. Kahane, eds., 191–212. Jerusalem: Academon (Hebrew).

Bean, F. D., L. B. Lowell, and L. J. Taylor. 1988. Undocumented Mexican immigrants and the earnings of other workers in the United States. *Demography* 25:35–49.

Beenstock, M., and Y. Ben-Menachem. 1995. *The labour market absorption of CIS immigrants to Israel: 1989–1994*. Jerusalem: Falk Institute for Economic Research (draft).

Beenstock, M., and Y. Fisher. 1996. *The contribution of aliya from the CIS to the Israeli economy in 1990–94: Final report*. Jerusalem: Si—Jerusalem Company for Research and Quantitative Analysis (Hebrew).

Ben-Bassat, A. 1991. Can Israel absorb half a million immigrants annually. *Economic Quarterly* 42(148): 267–274 (Hebrew).

Berry, B. J. L. 1988. Migration reversals in perspective: The long wave evidence. *International Regional Science Review* 11:245–251.

Birks, J. S., I. J. Seccombe, and C. A. Sinclair. 1988. Labor migration in the Arab Gulf states: Patterns, trends and prospects. *International Migration* 26:267–286.

Böhning, W. R. 1979. International migration in Western Europe: Reflections on the past five years. *International Labour Review* 118:401–414.

Böhning, W. R. 1984. *Studies in international labour migration.* London: Macmillan.

Borjas, G. J. 1983. The substitutability of black, Hispanic, and white labor. *Economic Inquiry* 21:93–106.

Borjas, G. J. 1985. Assimilation, changes in cohort quality, and the earnings of immigrants. *Journal of Labor Economics* 3:463–489.

Borjas, G. J. 1987a. Immigrants, minorities, and labor market competition. *Industrial and Labor Relations Review* 40:382–392.

Borjas, G. J. 1987b. Self-selection and the earnings of immigrants. *American Economic Review* 77:531–553.

Borjas, G. J. 1989. Economic theory and international migration. *International Migration Review* 23:457–485.

Borjas, G. J., and M. Tienda. 1987. The economic consequences of immigration. *Science* 235:643–651.

Borochov, E. 1991. Housing policy in view of the increase in immigration. *Economic Quarterly* 42(148): 49–60 (Hebrew).

Brochmann, G. 1993. Control in immigration policies: A close Europe in the making. In *The New Geography of European Migrations*, R. King, ed., 100–115. London: Belhaven Press.

Brutzkus, E. 1973. Changes in the network of urban centers in Israel. In *Towns in Israel*, A. Shachar, O. Weintraub, E. Cohen, and I. Shelack, eds., 155–168. Jerusalem: Academon (Hebrew).

Brown, L. A., K. V. L. England, and A. R. Goetz. 1989. Location, social categories, and individual labor market experiences in developing economies: The Venezuelan case. *International Regional Science Review* 12:1–28.

Brown, L. A., and A. R. Goetz. 1987. Development-related contextual effects and individual attributes in Third World migration processes: A Venezuelan example. *Demography* 24:497–516.

Brown, L. A., and V. A. Lawson. 1989. Polarization reversal, migration related shifts in human resource profiles, and spatial growth policies: A Venezuelan study. *International Regional Science Review* 12:185–188.

Bruno, M. 1990. *Economic strategy for growth and immigrant absorption.* Jerusalem: Bank of Israel (Hebrew).

Castles, S., and M. J. Miller. 1993. *The age of migration—international population movements in the modern world.* New York: Guilford Press.

Central Bureau of Statistics (CBS). 1993. *Immigration to Israel 1992*. Special publication series no. 944. Jerusalem: Central Bureau of Statistics.

Central Bureau of Statistics (CBS). 1995a. Employment of immigrants from USSR who arrived in Israel in October-December 1990: A follow-up survey, three years after immigration. *Monthly Bulletin of Statistics*, no. 6. Jerusalem: Central Bureau of Statistics.

Central Bureau of Statistics (CBS). 1995b. Households of immigrants from USSR who arrived in Israel in October-December 1990: A follow-up survey, three years after immigration. *Monthly Bulletin of Statistics*, no. 12. Jerusalem: Central Bureau of Statistics.

Central Bureau of Statistics (CBS). Selected years. *Construction in Israel*. Jerusalem: Central Bureau of Statistics.

Central Bureau of Statistics (CBS). Selected years. *Statistical Abstract of Israel*. Jerusalem: Central Bureau of Statistics.

Chalkley, B., and H. Winchester. 1991. Australia in transition. *Geography* 76:97–108.

Champion, A. G. 1988. The reversal of migration turnaround: Resumption of traditional trends? *International Regional Science Review* 11:253–260.

Champion, A. G. 1996. International migration and internal population movements within developed countries. Paper presented at the 28th International Geographical Congress, The Hague, August 4–10, 1996.

Chiswick, B. R. 1978. The effect of Americanization on the earnings of foreign-born men. *Journal of Political Economy* 86:897–921.

Chiswick, B. R. 1984. Illegal aliens in the United States labor market: Analysis of occupational attainment and earnings. *International Migration Review* 18:714–732.

Chiswick, B. R. 1986. Is the new immigration as unskilled as the old? *Journal of Labor Economics* 4:168–192.

Coates, B. E., R. J. Johnston, and P. L. Knox. 1977. *Geography and inequality*. Oxford: Oxford University Press.

Cochrane, S. G., and D. R. Vining, Jr. 1988. Recent trends in migration between core and peripheral regions in developed and advanced developing countries. *International Regional Science Review* 11:215–243.

Coleman, D. 1992. Does Europe need immigrants? Population and work force projections. *International Migration Review* 26:413–459.

Cruz, A. 1991. Will the European community make the 1992 deadline on the abolition of internal border controls? *International Migration* 24:477–482.

Damian, N., and Y. Rosenbaum-Tamari. 1996. *Immigrants from the CIS after five years in Israel (immigrants from July 1990)*. Report no. 9. Jerusalem: Ministry of Immigrant Absorption, Planning and Research Department (Hebrew).

Eckstein, Z. (chair). 1994. Final report of the committee appointed to examine plans for promoting the employment of immigrant engineers and other immigrant experts and for utilizing the human capital of the present wave of immigration (first draft). Presented to the Minister of Immigrant Absorption (Hebrew).

Eisenstadt, S. N. 1973. *Israeli society: Background, development, and problems*. Jerusalem: Magnes Press, Hebrew University (Hebrew).

Eisenstadt, S. N. 1976. Aliya and migration. In *The Social Structure of Israel*, S. N. Eisenstadt, H. Adler, R. Bar-Yosef, and R. Kahane, eds., 269–270. Jerusalem: Academon (Hebrew).

Enchautegui, M. E. 1992. Geographical differentials in the socioeconomic status of Puerto Ricans: Human capital variations and labor market characteristics. *International Migration Review* 26:1267–1289.

Felsenstein, D., and A. Shachar. 1986. The geography of the transit camps. *Idan (Immigrants and Transit Camps, 1948–1952)*, 87–96 (Hebrew).

Fielding, A. 1982. Counterurbanization in Western Europe. *Progress in Planning* 17, part 1: 1–52.

Fielding, A. J. 1993a. Mass migration and economic restructuring. In King 1993b, 7–18.

Fielding, A. J. 1993b. Migrants, institutions and politics: The evolution of European migration policies. In King 1993b, 40–62.

Filer, R. 1991. The effect of immigrant arrivals on migratory patterns of native workers. In *Immigration and the Work Force*, G. J. Borjas and R. B. Freeman, eds., 245–269. Chicago: University of Chicago Press.

Findlay, A. M. 1990. Migration channels approach to the study of high level manpower movements: A theoretical perspective. *International Migration* 28:15–23.

Findlay, A. M., and L. Garrick. 1990. Scottish emigration in the 1980s: A migration channels approach to study of skilled international migration. *Transactions, Institute of British Geographers* 13:177–192.

Findlay, A. M., F. L. N. Li, A. J. Jowett, and R. Skeldon. 1996. Skilled international migration and the global city: A study of expatriates in Hong Kong. *Transactions, Institute of British Geographers* 21:49–61.

Freeman, G. P. 1992. Migration policy and politics in the receiving states. *International Migration Review* 26:1144–1167.

Frey, W. H. 1995a. Immigration and internal migration 'flight': A California case study. *Population and Environment* 16:353–375.

Frey, W. H. 1995b. Immigration and internal migration flight from US metro areas: Toward a new demographic balkanization. *Urban Studies* 32:733–757.

Frey, W. H. 1996. Immigration, domestic migration, and demographic balkanization in America: New evidence for the 1990s. *Population and Development Review* 22:741–763.

Gonen, A. 1978. Urbanization of the moshavot in the land of Israel. *Studies in the Geography of Israel* 10:31–44 (Hebrew).

Gonen, A., ed. 1990. *The geography of immigrant absorption: Lessons of the past and a look to the future*. Jerusalem: Israel Geographical Association and the Hebrew University Department of Geography (Hebrew).

Gonen, A. 1991. Avoiding the mistakes of the past. *In the University*, journal of the Hebrew University of Jerusalem, pp. 27–30 (Hebrew).

Gordon, P. 1979. Deconcentration without a clean break. *Environment and Planning A* 11:281–290.

Gould, W. T. S., and A. M. Findlay. 1994. Population movements from the Third World to developed world: Recent trends and current issues. In *Population Migration and the Changing World Order*, A. M. Findlay and W. T. S. Gould, eds., 115–126. New York: John Wiley and Sons.

Gradus, Y. 1983. The role of politics in regional inequality: The Israeli case. *Annals of the Association of American Geographers* 73:388–403.

Gradus, Y. 1990. Population dispersion requires decentralization of government. In Gonen 1990, 34–35 (Hebrew).

Gradus, Y., and S. Krakover. 1976. The industrialization level and characteristics of industrialization in the central and development regions. *Economic Quarterly* 88–89:66–78 (Hebrew).

Gradus, Y., and S. Krakover. 1977. The effects of government policy on the spatial structure of manufacturing in Israel. *Journal of Developing Areas* 11:393–409.

Gradus, Y., and E. Stern. 1980. Changing strategies of development: Toward a regiopolis in the Negev Desert. *Journal of American Planning Association* 46:410–423.

Greenwood, M. J. 1994. Potential channels of immigrant influence on the economy of receiving country. *Papers in Regional Science* 73:211–240.

Greenwood, M. J., and G. L. Hunt. 1984. Migration and interregional employment redistribution in the United States. *American Economic Review* 74:957–969.

Greenwood, M. J., and J. M. McDowell. 1986. The factor market consequences of U.S. immigration. *Journal of Economic Literature* 24:1738–1772.

Grossman, J. B. 1982. The substitutability of natives and immigrants in production. *Review of Economics and Statistics* 64:596–603.

Hacohen, D. 1994. *Immigrants in turmoil: The great wave of immigration to Israel and its absorption, 1948–1953*. Jerusalem: Yad Ben-Zvi.

Hacohen, D. 1994. *The direct absorption method and its ramifications: The sociocultural absorption of the immigrants from the former Soviet Union in the early 1990s*. Jerusalem: Jerusalem Institute for Israel Studies (Hebrew).

Hall, P. 1993. Migration and the future of cities. Paper presented at European Migration and Urbanization, Umea, Sweden, June 1993.

Hanoch, G. 1961. *Income disparities in Israel*. Report no. 5, 1959/60. Jerusalem: Falk Institute (Hebrew).

Hasson, S. 1991. *Housing preferences of Soviet immigrants and their selection of place of residence: September 1989 to June 1990.* Jerusalem: Jerusalem Institute for Israel Studies, Urban Planning Division, and Ministry of Construction and Housing (Hebrew).

Hasson, S. 1993. *From international migration to internal migration: Selection of place of residence by immigrants from the former Soviet Union.* Jerusalem: Jerusalem Institute for Israel Studies and Ministry of Construction and Housing (Hebrew).

Herkovich, D., and Meridor, L. 1991. The macroeconomic implications of mass immigration to Israel. *Economic Quarterly* 42(148): 236–261 (Hebrew).

Hillard, P., D. Roche, A. Murie, and W. Birrel. 1972. Variation in the standard of housing provision in Northern Ireland. *Regional Studies* 6:393–399.

Hollifield, J. F., and G. Ross, eds. 1991. *Searching for the new France.* London: Routledge.

Jones, H. R. 1994. Immigration policy and the new world order: The case of Australia. In *Population Migration and the Changing World Order,* W. T. S. Gould and A. M. Findlay, eds., 161–172. New York: John Wiley and Sons.

Jones, R. C. 1995. Immigration reform and migrant flows: Compositional and spatial changes in Mexican migration after the Immigration Reform Act of 1986. *Annals of the Association of American Geographers* 85:715–730.

Kanjanapan, W. 1995. The immigration of Asian professionals to the United States: 1988–1990. *International Migration Review* 29:7–32.

Kellerman, A. 1986. Telecommunications as a tool for closing gaps between core and periphery. *Economic Quarterly* 128:547–554 (Hebrew).

Kellerman, A. 1993. *Society and settlement: Jewish Land of Israel in the twentieth century.* Albany, NY: State University of New York Press.

Kellerman, A., and S. Krakover. 1986. Multi-sectoral urban growth in space and time: An empirical approach. *Regional Studies* 20:117–129.

Kemper, F. J. 1993. New trends in mass migration in Germany. In King 1993b, 257–274.

Kemper, F. J. 1995. International migration to Germany: Recent developments in migration patterns and Germany's pivotal role within new European migration systems. In *Japan and Central Europe Restructuring: Geographical Aspects of Socio-economic, Urban and Regional Development,* W. Flüchter, ed., 201–210. Wiesbaden: Harrassowitz.

Kim, I. 1987. The Koreans: Small business in an urban frontier. In *New Immigrants in New York,* N. Foner, ed., 219–242. New York: Columbia University Press.

King, R. 1991. Europe's metamorphic demographic map. *Town and Country Planning* 60:111–113.

King, R. 1993a. European international migration 1945–1990: A statistical and geographical overview. In King 1993b, 19–39.

King, R., ed., 1993b. *Mass migration in Europe: The legacy and the future.* London: Belhaven Press.

King, R. 1993c. Why do people migrate? The geography of departure. In *The New Geography of European Migrations*, R. King, ed., 17–46. London: Belhaven Press.

Kipnis, B. A. 1983. The development of the Jewish towns in the Galilee 1948–1980. In *The Lands of the Galilee*, A. Shmueli, A. Soffer, and N. Kliot, eds., 723–744. Haifa: University of Haifa and Ministry of Defense.

Kipnis, B. A. 1984. Role and timing of complementary objectives of a regional policy: The case of northern Israel. *Geoforum* 15:191–200.

Kipnis, B. A. 1989. *Plant scale and manufacturing policies for peripheral regions.* Aldershot: Avebury.

Kipnis, B. A. 1990. Have we abolished the population dispersion policy? In Gonen 1990, 17–28 (Hebrew).

Klassen, L. H., and J. H. Paelinck. 1979. The future of large towns. *Environment and Planning A* 11:1091–1104.

Krakover, S., and E. Stern. 1990. Application of geographical models to absorbing the wave of immigration from the USSR. In Gonen 1990, 5–16 (Hebrew).

Krätke, S. 1989. The future of social housing: Problems and prospects of social ownership: The case of West Germany. *International Journal of Urban and Regional Research* 13:282–303.

Kuz, T. J. 1978. Quality of life, an objective and subjective variable analysis. *Regional Studies* 12:409–417.

Kuznets, S. 1960. Population change and aggregate output. In *Demographic and Economic Change in Developed Countries*, A. J. Coale, ed., 324–351. Princeton: Princeton University Press.

Lavi, V. 1988. *Unemployment in development towns.* Study no. 29. Jerusalem: Jerusalem Institute for Israel Studies (Hebrew).

Law-Yone, U., and G. Lipshitz. 1991. Goals, ambiguity and adhocism—the new settlements program in the Galilee. Haifa: Center for Urban and Regional Research, Technion.

Lehrman, A., R. Lehrman, and National Planning and Building Board (NPBB). 1992. *Combined national outline plan for building and development and immigrant absorption, NOP 31.* Jerusalem: National Planning and Building Board (Hebrew).

Levitan, N. 1964. *Consumption patterns in Israel.* Falk Institute (Hebrew).

Lipshitz, G. 1986a. A spatial typology of Israel on the basis of partial order analysis. *Geographical Analysis* 18:324–342.

Lipshitz, G. 1986b. The stability of the spatial pattern of welfare, the Israeli case. *Geoforum* 17:353–366.

Lipshitz, G. 1987. Migration within metropolitan areas of Israel, comparative periodic analysis. *Papers of Regional Science Association* 61:161–174.

Lipshitz, G. 1990. *Development towns: A new basis for policy planning.* Jerusalem: Jerusalem Institute for Israel Studies (Hebrew).

Lipshitz, G. 1991a. Immigration and internal migration as a mechanism of polarization and dispersion of population and development, the Israeli case. *Economic Development and Cultural Change* 39:391–408.

Lipshitz, G. 1991b. Ethnic differences in migration patterns—Disparities among Arabs and Jews in the peripheral regions of Israel. *Professional Geographer* 49:445–456.

Lipshitz, G. 1992a. *Development towns that take in immigrants: Dimona and Qiryat Shemona*. Jerusalem: Jerusalem Institute for Israel Studies.

Lipshitz, G. 1992b. Divergence versus convergence in regional development—the evolution of the theoretical thought. *Journal of Planning Literature* 7:123–138.

Lipshitz, G. 1993a. Dispersal of high-tech settlements as a strategy for peripheral regional development. *TESG—Journal of Economic and Social Geography* 84:40–50.

Lipshitz, G. 1993b. The main approaches to measuring regional development and welfare. *Social Indicators Research—an International and Interdisciplinary Journal for Quality-of-Life* 29:161–181.

Lipshitz, G. 1996a. *Regional disparities, immigration, and internal migration: The Canadian case in the theoretical context*. Jerusalem: Hebrew University of Jerusalem, Center for Canadian Studies.

Lipshitz, G. 1996b. Spatial concentration and deconcentration of population: Israel as a case study. *Geoforum* 27:87–96.

Lipshitz, G. 1997. Immigrants from the former Soviet Union in the Israeli housing market: Spatial aspects of supply and demand. *Urban Studies* 34:471–488.

Lithwick, E., and G. Noam. 1995. Absorption of immigrants from the former Soviet Union in Sederot: Census findings. Jerusalem: JDC–Brookdale Institute (Hebrew).

Massey, D. S. 1984. *Spatial divisions of labour*. London: Macmillan.

Massey, D. S. 1987. Do undocumented migrants earn lower wages than legal immigrants? New evidence from Mexico. *International Migration Review* 21:236–274.

Meckie, C., and K. Thompson. 1990. *Canadian social trends*, Toronto: Thompson Educational Publishers.

Mera, K. 1988. The emergence of migration cycles? *International Regional Science Review* 11:269–275.

Min, P. G. 1984. From white-collar occupations to small business: Korean immigrants' occupational adjustment. *Sociological Quarterly* 25:333–352.

Ministry of Immigrant Absorption. 1991. *Report on immigrants from the Soviet Union, 1990 January 1990–December 1990): Statistical data*. Produced by Malam Systems, Ltd. Jerusalem: Ministry of Immigrant Absorption, Information Services Unit (Hebrew).

Ministry of Immigrant Absorption. 1995. *Immigrant absorption: The present situation, challenges, and goals* Jerusalem: Ministry of Immigrant Absorption (Hebrew).

Moore, E. G., B. K. Ray, and M. W. Rosenberg. 1990. The redistribution of immigrants in Canada. Population working paper no. 12. In *Employment and Immigration*

Canada. Ottawa: Hull.

Morrison, P. A., and J. P. Wheeler. 1976. Rural renaissance in America. *Population Bulletin* 31:1–27.

Myrdal, G. 1957. *Economic theory and underdeveloped regions*. London: Duckworth.

Nirel, N., and G. Naveh. 1996. *The employment of immigrant physicians in Israel: Is it stable? Selected characteristics of the employment of immigrant physicians from the former Soviet Union*. Jerusalem: JDC–Brookdale Institute of Gerontology and Human Development (Hebrew).

Naveh, G., M. Noam, and E. Benita. 1993. *The employment and economic status of the immigrants: Selected findings from the national employment survey of Soviet immigrants*. Jerusalem: JDC–Brookdale Institute (Hebrew).

Newman, D., Y. Gradus, E. Levinson, and E. Kalati. 1993. *The municipal functioning of northern Negev towns following the immigration from the former Soviet Union*. Beersheva: Ben-Gurion University of the Negev and Negev Center for Regional Development, for the Ministry of Construction and Housing (Hebrew).

Nijkamp, P., and K. Spiess. 1993. International migration in Western Europe: Macro trends in the past, the present and the future. *Current Politics and Economics of Europe* 3:238–262.

Nijkamp, P., and K. Spiess. 1994a. International migration in Europe: Critical success absorption factors. *Journal of Regional Studies* 12:331–335.

Nijkamp, P., and K. Spiess. 1994b. Missing fields in regional science research: European and global views on a mobile network society. Paper presented at RSA—International Conference, Moscow.

Nogle, J. M. 1994. Internal migration for recent immigrants to Canada. *International Migration Review* 28:31–48.

Öberg, S., and H. Boubnova. 1993. Ethnicity, nationality and migration potentials in Eastern Europe. In King 1993b, 234–256.

Ofer, G., K. Plug, and N. Kasir. 1991. Employment of immigrants from the Soviet Union in 1990 and later: Occupational stability and switching. *Economic Quarterly* 42(148): 135–179 (Hebrew).

Ozdowski, S. A. 1985. The law, immigration and human rights: Changing the Australian control system. *International Migration Review* 19:535–554.

Patinkin, D. 1959. *The Israeli economy in the first decade*. Report no. 4, 1957/1958. Jerusalem: Falk Institute (Hebrew).

Patterson, R. 1990. Development of ethnic multicultural media in Australia. *International Migration* 28:89–103.

Pekin, H. 1989. Effect of migration in family structure. *International Migration Review* 27:281–293.

Piore, M. J. 1979: *Birds of passage: Migrant labor and industrial societies*. New York: Cambridge University Press.

Portes, A., and R. G. Rumbaut. 1990. *Immigrant America.* Berkeley: University of California Press.

Poston, D. L. 1994. Patterns of economic attainment of foreign-born male workers in the United States. *International Migration Review* 28:478–499.

Razin, A. and E. Sadka. 1995. *Population economics*, chap. 16. Cambridge, Mass.: MIT Press.

Reichmann, S. 1973. On the validity of the spatial distribution plans of Israel's population. *City and Region* 1:26–43 (Hebrew).

Reichmann, S. 1979. *From foothold to settled territory: Development of the settlement map in Palestine, 1918–1948.* Jerusalem: Yad Ben-Zvi (Hebrew).

Reichmann, S., A. Eliash, and D. Berashi. 1971. *Travel patterns in Israel 1958–1971.* Tel Aviv: Israeli Institute for Transportation Research and Planning (Hebrew).

Richmond, A. H. 1992. Immigration and structural change: The Canadian experience, 1971–1986. *International Migration Review* 26:1200–1221.

Salt, J. 1984. High-level manpower movement in North-West Europe and the role of careers. *International Migration Review* 17:633–651.

Salt, J. 1989. A comparative overview of international trends and types, 1950–1980. *International Migration Review* 23:431–456.

Salt, J. 1992a. Migration processes among the highly skilled in Europe. *International Migration Review* 26:484–505.

Salt, J. 1992b. The future of international labor migration. *International Migration Review* 26:1077–1143.

Salt, J., and R. Ford. 1993. Skilled international migration in Europe: The shape of things to come? In King 1993b, 293–309.

Sassen, S. 1988. *The mobility of labour and capital.* Cambridge: Cambridge University Press.

Sassen, S. 1993. Economic internationalization: The new migration in Japan and the United States. *International Migration* 31:73–102.

Segev, T. 1984. *The first Israelis—1949.* Jerusalem: Domino Press.

Shachar, A. 1971. Israel's development towns: Evaluation of national urban policy. *Journal of the American Institute of Planners* 17:271–291.

Shachar, A. 1975. Patterns of population densities in the Tel-Aviv metropolitan area. *Environment and Planning A* 7:271–291.

Shachar, A., and G. Lipshitz. 1980. The spatial structure of interregional migration in Israel. *Studies in the Geography of Israel* 11:153–178 (Hebrew).

Shachar, A., and G. Lipshitz. 1981. Regional inequalities in Israel. *Environment and Planning A* 13:373–463.

Shachar, A., and S. Öberg. 1990. *The world economy and the spatial organization of power.* Aldershot, England: Avebury.

Sharon, A. 1973. Elements of physical planning in Israel. In *Towns in Israel*, A. Shachar,

O. Weintraub, E. Cohen, and I. Shelack, eds., 126–141. Jerusalem: Academon (Hebrew).

Shefer, D. 1991. The housing crisis and immigrant absorption. *Immigrant Absorption, Building, and National Planning* (proceedings of the annual conference of the Environmental Planning Society), 7–12. Haifa: Center for Urban and Regional Research, the Technion (Hebrew).

Shefer, D., and A. Frenkel. 1989. *Job creation in development towns*. Haifa: The Technion, S. Neaman Institute, Center for Urban and Regional Research (Hebrew).

Sheinin, Y. 1991. Building housing for one million immigrants. *Economic Quarterly* 42(148): 39–48 (Hebrew).

Shevtsova, L. 1992. Post Soviet emigration today and tomorrow. *International Migration Review* 26:241–257.

Shin, E. H. and K. S. Chang. 1988. Peripherization of immigrant professionals: Korean physicians in the United States. *International Migration Review* 22:81–84.

Sicron, M. 1957. *Immigration to Israel, 1949–1953*. Jerusalem: Falk Institute and Central Bureau of Statistics (Hebrew).

Sicron, M., and B. Lassman. 1976. Changes in the population of districts in Israel, 1961–1972. *City and Region* 1–2:23–24 (Hebrew).

Simon, J. L. 1992. *The economic consequences of immigration*. London: Routledge.

Slater, D. 1975. Underdevelopment and spatial inequality. *Progress in Planning* 4:101–165.

Smith, D. M. 1979. *Where the grass is greener: Living in an unequal world.* Hammondsworth, England: Penguin Books.

Smith, D. M. 1994. *Geography and social justice*. Oxford: Blackwell.

Soffer, A., and R. Finkel. 1988. *The mitzpim in the Galilee—goals and results*. Rehovot, Israel: Center for Rural and Urban Settlements.

Sonis, M. 1980. Locational push-pull analysis of migration streams. *Geographical Analysis* 12:80–97.

Sonis, M. 1988. Interregional migration—Israel. In *Interrregional Migration: Dynamic Theory and Comparative Analysis*, W. Weidlich and G. Haag, eds., 155–184. Berlin: Springer.

Sporton, D., and P. E. White. 1989. Immigrants in social housing: Integration or segregation in France? *The Planner* 75(4): 2–31.

Stahl, C. W., and R. T. Appleyard. 1992. International manpower flows in Asia: An overview. *Asian and Pacific Migration Journal* (special issue) 1:417–476.

Stern, E. 1988. The spatial implication of the tax policy on the private motorization rate in Israel. *City and Region* 18:48–64 (Hebrew).

Stern, E., and S. Krakover. 1994. *Expedited building sites*. Jerusalem: Planning Environments, Ministry of Construction and Housing (Hebrew).

Szoke, L. 1992. Hungarian perspectives on emigration and immigration in the new

European architecture. *International Migration Review* 26:305–324.

Usher, D. 1977. Public property and the effects of migration upon other residents of the migrants' countries of origin and destination. *Journal of Political Economy* 85:1001–1020.

Van Amersfoort, H., and L. de Klerk. 1987. Dynamics of immigrant settlement: Surinamese, Turks and Moroccans in Amsterdam, 1973–1983. In *Foreign Minorities in Continental European Cities*, G. Glebe and J. O'Loughlin, eds., 199–222. Wiesbaden: Steiner.

Vining, D. R., Jr., and T. P. Kontuly. 1978. Population dispersion from major metropolitan regions: An international comparison. *International Regional Science Review* 2:49–73.

Vining, D. R., Jr., and R. Pallone. 1982. Migration between core and periphery regions: A description and tentative explanation of the patterns of 22 countries. *Geoforum* 13:339–410.

Vining, D. R., Jr., and A. Strauss. 1977. A demonstration that the current deconcentration of population in the United States is a clean break with the past. *Environment and Planning A* 9:751–758.

Wallace, S. P. 1986. Central American and Mexican immigrant characteristics and economic incorporation in California. *International Migration Review* 20:657–671.

Weingrod, E., and M. Minkowicz. 1963. Patterns of absorption of the second generation on immigrant moshavim. *Trends* 124:361–372 (Hebrew).

Weintraub, D. 1954. Absorption problems and integration on immigrant moshavim. *Trends* 5(3): 248–260 (Hebrew).

Weintraub, D. 1960. The impact of the demographic makeup of immigrant families on their adjustment on the moshav. *Trends* 10(4): 312–321 (Hebrew).

Weitz, R. 1957. Seven years of new settlement. *Economic Quarterly* 3(11): 269–275 (Hebrew).

Weitz, R., and A. Rokeach. 1986. Mass aliya and rural settlement projects. In *Idan (Immigrants and Transit Camps, 1948–1952)*, 53–68 (Hebrew).

White, P. 1993. The social geography of immigrants in European cities: The geography of arrival. In *The New Geography of European Migrations*, R. King, ed., 47–65. London: Belhaven Press.

Widgren, J. 1990. International migration and regional stability. *International Affairs* 66:749–766.

World Bank. 1993. *The East Asian miracle. Economic growth and public policy.* Oxford University Press.

Zilberger, R. 1973. *Population distribution in Israel, 1948–1972.* Jerusalem: Ministry of Finance, Economic Planning Authority (Hebrew).

Zolberg, A. R. 1989. The next waves: Migration theory for a changing world. *International Migration Review* 23:403–430.

INDEX

The GeoJournal Library

1. B. Currey and G. Hugo (eds.): *Famine as Geographical Phenomenon.* 1984
ISBN 90-277-1762-1
2. S.H.U. Bowie, F.R.S. and I. Thornton (eds.): *Environmental Geochemistry and Health.* Report of the Royal Society's British National Committee for Problems of the Environment. 1985 ISBN 90-277-1879-2
3. L.A. Kosiński and K.M. Elahi (eds.): *Population Redistribution and Development in South Asia.* 1985 ISBN 90-277-1938-1
4. Y. Gradus (ed.): *Desert Development.* Man and Technology in Sparselands. 1985 ISBN 90-277-2043-6
5. F.J. Calzonetti and B.D. Solomon (eds.): *Geographical Dimensions of Energy.* 1985 ISBN 90-277-2061-4
6. J. Lundqvist, U. Lohm and M. Falkenmark (eds.): *Strategies for River Basin Management.* Environmental Integration of Land and Water in River Basin. 1985 ISBN 90-277-2111-4
7. A. Rogers and F.J. Willekens (eds.): *Migration and Settlement.* A Multiregional Comparative Study. 1986 ISBN 90-277-2119-X
8. R. Laulajainen: *Spatial Strategies in Retailing.* 1987 ISBN 90-277-2595-0
9. T.H. Lee, H.R. Linden, D.A. Dreyfus and T. Vasko (eds.): *The Methane Age.* 1988 ISBN 90-277-2745-7
10. H.J. Walker (ed.): *Artificial Structures and Shorelines.* 1988
ISBN 90-277-2746-5
11. A. Kellerman: *Time, Space, and Society.* Geographical Societal Perspectives. 1989 ISBN 0-7923-0123-4
12. P. Fabbri (ed.): *Recreational Uses of Coastal Areas.* A Research Project of the Commission on the Coastal Environment, International Geographical Union. 1990 ISBN 0-7923-0279-6
13. L.M. Brush, M.G. Wolman and Huang Bing-Wei (eds.): *Taming the Yellow River: Silt and Floods.* Proceedings of a Bilateral Seminar on Problems in the Lower Reaches of the Yellow River, China. 1989 ISBN 0-7923-0416-0
14. J. Stillwell and H.J. Scholten (eds.): *Contemporary Research in Population Geography.* A Comparison of the United Kingdom and the Netherlands. 1990
ISBN 0-7923-0431-4
15. M.S. Kenzer (ed.): *Applied Geography.* Issues, Questions, and Concerns. 1989 ISBN 0-7923-0438-1
16. D. Nir: *Region as a Socio-environmental System.* An Introduction to a Systemic Regional Geography. 1990 ISBN 0-7923-0516-7
17. H.J. Scholten and J.C.H. Stillwell (eds.): *Geographical Information Systems for Urban and Regional Planning.* 1990 ISBN 0-7923-0793-3
18. F.M. Brouwer, A.J. Thomas and M.J. Chadwick (eds.): *Land Use Changes in Europe.* Processes of Change, Environmental Transformations and Future Patterns. 1991 ISBN 0-7923-1099-3
19. C.J. Campbell: *The Golden Century of Oil 1950–2050.* The Depletion of a Resource. 1991 ISBN 0-7923-1442-5
20. F.M. Dieleman and S. Musterd (eds.): *The Randstad: A Research and Policy Laboratory.* 1992 ISBN 0-7923-1649-5
21. V.I. Ilyichev and V.V. Anikiev (eds.): *Oceanic and Anthropogenic Controls of Life in the Pacific Ocean.* 1992 ISBN 0-7923-1854-4

The GeoJournal Library

22. A.K. Dutt and F.J. Costa (eds.): *Perspectives on Planning and Urban Development in Belgium.* 1992 ISBN 0-7923-1885-4
23. J. Portugali: *Implicate Relations.* Society and Space in the Israeli-Palestinian Conflict. 1993 ISBN 0-7923-1886-2
24. M.J.C. de Lepper, H.J. Scholten and R.M. Stern (eds.): *The Added Value of Geographical Information Systems in Public and Environmental Health.* 1995 ISBN 0-7923-1887-0
25. J.P. Dorian, P.A. Minakir and V.T. Borisovich (eds.): *CIS Energy and Minerals Development.* Prospects, Problems and Opportunities for International Cooperation. 1993 ISBN 0-7923-2323-8
26. P.P. Wong (ed.): *Tourism vs Environment: The Case for Coastal Areas.* 1993 ISBN 0-7923-2404-8
27. G.B. Benko and U. Strohmayer (eds.): *Geography, History and Social Sciences.* 1995 ISBN 0-7923-2543-5
28. A. Faludi and A. der Valk: *Rule and Order. Dutch Planning Doctrine in the Twentieth Century.* 1994 ISBN 0-7923-2619-9
29. B.C. Hewitson and R.G. Crane (eds.): *Neural Nets: Applications in Geography.* 1994 ISBN 0-7923-2746-2
30. A.K. Dutt, F.J. Costa, S. Aggarwal and A.G. Noble (eds.): *The Asian City: Processes of Development, Characteristics and Planning.* 1994 ISBN 0-7923-3135-4
31. R. Laulajainen and H.A. Stafford: *Corporate Geography.* Business Location Principles and Cases. 1995 ISBN 0-7923-3326-8
32. J. Portugali (ed.): *The Construction of Cognitive Maps.* 1996 ISBN 0-7923-3949-5
33. E. Biagini: *Northern Ireland and Beyond.* Social and Geographical Issues. 1996 ISBN 0-7923-4046-9
34. A.K. Dutt (ed.): *Southeast Asia: A Ten Nation Region.* 1996 ISBN 0-7923-4171-6
35. J. Settele, C. Margules, P. Poschlod and K. Henle (eds.): *Species Survival in Fragmented Landscapes.* 1996 ISBN 0-7923-4239-9
36. M. Yoshino, M. Domrös, A. Douguédroit, J. Paszynski and L.D. Nkemdirim (eds.): *Climates and Societies – A Climatological Perspective.* A Contribution on Global Change and Related Problems Prepared by the Commission on Climatology of the International Geographical Union. 1997 ISBN 0-7923-4324-7
37. D. Borri, A. Khakee and C. Lacirignola (eds.): *Evaluating Theory-Practice and Urban-Rural Interplay in Planning.* 1997 ISBN 0-7923-4326-3
38. J.A.A. Jones, C. Liu, M-K. Woo and H-T. Kung (eds.): *Regional Hydrological Response to Climate Change.* 1996 ISBN 0-7923-4329-8
39. R. Lloyd: *Spatial Cognition.* Geographic Environments. 1997 ISBN 0-7923-4375-1
40. I. Lyons Murphy: *The Danube: A River Basin in Transition.* 1997 ISBN 0-7923-4558-4
41. H.J. Bruins and H. Lithwick (eds.): *The Arid Frontier.* Interactive Management of Environment and Development. 1998 ISBN 0-7923-4227-5

The GeoJournal Library

42. G. Lipshitz: *Country on the Move: Migration to and within Israel, 1948–1995.* 1998 ISBN 0-7923-4850-8
43. S. Musterd, W. Ostendorf and M. Breebaart: *Multi-Ethnic Metropolis: Patterns and Policies.* 1998 ISBN 0-7923-4854-0
44. B.K. Maloney (ed.): *Human Activities and the Tropical Rainforest.* Past, Present and Possible Future. 1998 ISBN 0-7923-4858-3
45. H. van der Wusten (ed.): *The Urban University and its Identity.* Roots, Location, Roles. 1998 ISBN 0-7923-4870-2

KLUWER ACADEMIC PUBLISHERS – DORDRECHT / BOSTON / LONDON

GPSR Compliance

The European Union's (EU) General Product Safety Regulation (GPSR) is a set of rules that requires consumer products to be safe and our obligations to ensure this.

If you have any concerns about our products, you can contact us on ProductSafety@springernature.com

In case Publisher is established outside the EU, the EU authorized representative is:

Springer Nature Customer Service Center GmbH
Europaplatz 3
69115 Heidelberg, Germany

Batch number: 10823959

Printed by Printforce, the Netherlands